The Aquarium Fish Medicine Handbook

Practicing veterinarians, veterinary technicians, professional aquarists, fish researchers, and tropical fish hobbyists will find this thorough yet concise handbook a complete how-to guide for keeping fish under human care healthy and thriving. Whether you're new to fish medicine and husbandry or an expert in the field, there is valuable information within these pages for you. The book is arranged in a logical order so that readers can quickly and efficiently find the information they seek. The book is filled with pertinent and applicable facts on dozens of topics, including how to manage aquatic life support systems for fish, history taking, natural history, anatomy, physiology, diagnostic techniques, anesthesia, analgesia, and surgery. Virtually all the most important ornamental fish pathogens, including viruses, bacteria, fungi, and parasites, are described and richly illustrated. There are also sections on fish welfare and conservation.

While there are many books on the market that explore the topics of ornamental fish health and care, none provide the amount of helpful information and resources in such a small, affordable package.

The Aquarium Fish Medicine Handbook

**ROY P.E. YANONG
AND GREGORY A. LEWBART**

CRC Press
Taylor & Francis Group
Boca Raton London New York

CRC Press is an imprint of the
Taylor & Francis Group, an **informa** business

Cover credit: The Oranda Red Cap Goldfish was courtesy of 5D Tropical, Inc., and the Copperband Butterfly was courtesy of Segrest, Inc.

First edition published 2024
by CRC Press
2385 NW Executive Center Drive, Suite 320, Boca Raton, FL 33431

and by CRC Press
4 Park Square, Milton Park, Abingdon, Oxon, OX14 4RN

CRC Press is an imprint of Taylor & Francis Group, LLC

©2024 Roy P.E. Yanong and Gregory A. Lewbart

Reasonable efforts have been made to publish reliable data and information, but the author and publisher cannot assume responsibility for the validity of all materials or the consequences of their use. The authors and publishers have attempted to trace the copyright holders of all material reproduced in this publication and apologize to copyright holders if permission to publish in this form has not been obtained. If any copyright material has not been acknowledged please write and let us know so we may rectify in any future reprint.

Except as permitted under U.S. Copyright Law, no part of this book may be reprinted, reproduced, transmitted, or utilized in any form by any electronic, mechanical, or other means, now known or hereafter invented, including photocopying, microfilming, and recording, or in any information storage or retrieval system, without written permission from the publishers.

For permission to photocopy or use material electronically from this work, access www.copyright.com or contact the Copyright Clearance Center, Inc. (CCC), 222 Rosewood Drive, Danvers, MA 01923, 978-750-8400. For works that are not available on CCC please contact mpkbookspermissions@tandf.co.uk

Trademark notice: Product or corporate names may be trademarks or registered trademarks and are used only for identification and explanation without intent to infringe.

ISBN: 9781032711201 (hbk)
ISBN: 9780367522919 (pbk)
ISBN: 9781003057727 (ebk)

DOI: 10.1201/9781003057727

Typeset in Janson
by Deanta Global Publishing Services, Chennai, India

To our parents:

the late Virginia and Marvin Lewbart, who nurtured my curiosity of nature, supported my ideas and efforts, and loved me unconditionally for over 64 years,

and

Gregoria and Procopio Yanong, proud Filipino immigrants and compassionate physicians, who taught me the importance of family, integrity, perseverance, empathy, and service by their sacrifices, support, and unconditional love.

Gregory Lewbart and Roy Yanong

TABLE OF CONTENTS

vii

About the Authors		xi
Introduction		xii
Acknowledgements		xiii
CHAPTER 1	**AQUARIUM FISH FAMILIES AND SPECIES**	1
	Common Groups Kept	1
	Popular Freshwater Ornamental Fishes	1
	Popular Pond Fishes	18
	Popular Tropical Marine Ornamental Fishes	18
	Popular Aquarium Non-Fish Species	34
	Laws and Permits	35
CHAPTER 2	**AQUARIUM SYSTEMS AND WATER QUALITY**	36
	Important System Processes	36
	Tank Essentials	42
	Water Quality: Chemical Imbalances And Management	55
	Tank And Water Tips	71
	Salt Calculations for Common Volumes of Water	75
CHAPTER 3	**NUTRITION AND DIET**	77
	General Nutrient Requirements and Feed Ingredients	77
	Macronutrients	77
	Micronutrients/Minerals	78
	Additional Considerations	78
	Amount	81

CHAPTER 4	**GENERAL REPRODUCTION**	85
	Sexing Select Species/Groups	85
	General Reproductive Strategies	89
	Common Marine Species	91
	General Notes on Tropical Fish Early Life Stages	92
	Egg Binding (Dystocia)	92
CHAPTER 5	**CLINICAL CASE WORK UPS: INITIAL CONSIDERATIONS**	95
	First Visit/Annual Examination Protocol	95
	Fish Physical Examination	97
	Diagnostics	98
	Emergency Therapy/First Aid	100
CHAPTER 6	**ANTE-MORTEM DIAGNOSTICS**	102
	External Wet Mount Biopsies	102
	Bacterial Culture – Non-Sacrificial	102
	Blood Collection Techniques	107
	Diagnostic Evaluation of Samples	113
	Hematology	113
CHAPTER 7	**IMAGING**	123
	Radiography And Ultrasonography: General Comments	123
CHAPTER 8	**POST-MORTEM EVALUATION**	127
	General Comments on Specimen Quality	127
	Necropsy Supplies	128
	Microbiology Supplies	128
	Wet Mount Evaluation	129
	Necropsy and Microbiology	133
	Internal Evaluation	136

CHAPTER 9	**EUTHANASIA, ANESTHESIA, AND SEDATION**	149
	General Considerations for Euthanasia	149
	Indicators of Death in Fish	149
	Euthanasia Methods	149
	Anesthesia And Sedation	151
CHAPTER 10	**SURGERY**	153
	Common Indications for Surgery	153
	Fish Anesthesia Delivery System (FADS)	153
	Induction, Maintenance, Monitoring	155
	Fish Preparation	155
	Surgical Tips	157
CHAPTER 11	**INFECTIOUS DISEASES**	159
	General Management Considerations	159
	Preventive Care: Vaccines	160
	Parasitic Diseases of Aquarium Fish	160
	Bacterial Diseases of Aquarium Fish	197
	Viral Diseases of Aquarium Fish	205
	Fungal Diseases of Fish	214
	Reportable Diseases	219
	Zoonoses	220
CHAPTER 12	**COMMON MEDICAL CONDITIONS**	222
	Fatty Liver Disease	222
	Neoplasia	222
	Buoyancy Problems (Swim Bladder "Disease")	222
	Superficial Wounds	224
	Signs Associated with Water Quality Problems	224
	Lateral Line Depigmentation Syndrome (LLD) of Marine Fish	225
	Ocular Diseases	226

	Nutritional Diseases	227
	Constipation	228
	Behavioral Considerations	228
CHAPTER 13	HEALTH CERTIFICATES/CERTIFICATES OF VETERINARY INSPECTION	230
CHAPTER 14	DRUGS AND CHEMICALS: ADMINISTRATION METHODS AND FORMULARY	232
	Pre-Treatment Considerations	232
	Routes of Drug and Chemical Administration for Ornamental Fish	232
	Formulary	239
	Drugs and Dosages	240
	Bibliography	254
CHAPTER 15	EQUIPMENT AND SUPPLIES CHECKLIST	262
	Husbandry	262
	Medical Supplies	262
	Microbiology	263
	Drugs and Other Compounds	263
	Diagnostic Supplies	265
CHAPTER 16	MATERIAL RESOURCES	267
	Products and Services	267
	Water Chemistry Test Kits	268
	Fish Drug Distributors	269
	Other Informational Resources	269

References/Further Reading — 271

Index — 275

ABOUT THE AUTHORS

Roy Yanong worked for a large ornamental fish farm before joining the University of Florida IFAS Tropical Aquaculture Laboratory in 1996. As Professor and Extension Veterinarian, he provides outreach, research, and educational programs in disease diagnostics and health management. He has authored numerous articles and book chapters, co-founded the American Association of Fish Veterinarians, and hosts the Aquariumania podcast on PetLifeRadio.

Greg Lewbart worked for a large wholesaler of ornamental fishes before joining the faculty at the North Carolina State College of Veterinary Medicine in 1993, where he is Professor of Aquatic, Wildlife, and Zoological Medicine. He is a diplomate of the American and European Colleges of Zoological Medicine and has authored more than 200 scientific articles, 35 book chapters, and edited or co-edited five veterinary textbooks.

INTRODUCTION

FISH

The information presented here has been compiled from the literature and the authors' clinical experience. It is intended to be used as a quick guide to selected husbandry and medical topics of fish and is not intended to replace more detailed reference material.

ACKNOWLEDGEMENTS

We would first like to acknowledge the individuals who have been a source of guidance, inspiration, and support over many years. We thank all our mentors and professors, but in particular, Donald Abt – original builder and captain of the grand ship *AQUAVET* – and first mate Paul Bowser, Robert Barnes, Philip Bookman, Joe Diaz, Dale Dickey, John Gratzek, Louis Leibovitz, William Medway, Edward Noga, David O'Beirne, Nathan "Doc" Riser, Stephen Smith, Roxanna Smolowitz, Ralph Sorensen, Michael Stoskopf, Jeff Wolf, Brent Whitaker, and Richard Wolke.

I am fortunate to be associated with the North Carolina State University College of Veterinary Medicine (NC State CVM), an inclusive, progressive institution of higher learning. I am grateful to my NC State friends and colleagues. Clarke Atkins, Anthony Blikslager, John Cullen, Elizabeth Hardie, Paul Lunn, Kate Meurs, Kent Passingham, Elizabeth Stone, and Michael Stoskopf have been especially supportive. I particularly cherish my over three-decade friendship with Craig Harms. I am also extremely grateful to my close friend, respected colleague, international stalwart in the field of ornamental fish medicine, and lead author on this project, Roy Yanong.

An ocean of gratitude to my great mentor, friend, and co-author Greg Lewbart who gave me a compass and then set my professional life on its course. Thanks to all my University of Florida (UF) colleagues, but especially to those at the Tropical Aquaculture Laboratory and the College of Veterinary Medicine, for fostering synergy, productivity, and fun. Craig Watson and Ruth Francis-Floyd taught me how to live UF's land grant mission; Debbie Pouder and Eric Curtis partnered with me to live it – assisting clients and educating students. Heartfelt thanks also go to Ilze Berzins, Hugh Mitchell, and Johnny Shelley for helping me navigate tricky channels en route to major aquatic landmarks. Finally, my appreciation to the Penn Vet Class of 1992, especially to members of Hyrax House, for their ongoing support, friendship, humor, and love.

We collectively thank the graduate and veterinary students, interns, residents, and house officers we have worked with, both at NC State and UF, and those from other veterinary colleges. On many days, these young people teach us more than we teach them, and they are the bright future and promise of our profession.

Thanks to Greg Trende for his assistance with this book and to all our colleagues who graciously contributed figures, including Kirstin Cook, who provided beautiful anatomical illustrations.

The folks at CRC Press have been exceptional through this entire process. We specifically acknowledge Alice Oven, Amelia Bashford, Shikha Garg, Marsha Hecht, and Jayanthi Chander. Jayanthi, the book's project manager, was especially patient and attentive. A special thank you to Linda and Greg Harrison of the Zoological Education Network for the idea of this handbook and for their continued generosity and support.

Finally, and most importantly, I am grateful for the love, support, and wise insight provided by my wife, life partner, and confidant, Dr. Diane Deresienski.

CHAPTER 1
AQUARIUM FISH FAMILIES AND SPECIES

COMMON GROUPS KEPT

- At least 4,000 species of fish are kept as pets or in aquariums (not including morphological, color, and pattern varieties). There are about 100 commonly kept species.
- Most pet fish fall into five categories:
    1. Tropical freshwater species, such as tetras, algae eaters, silver dollars, danios, mollies, swords, barbs, cichlids, and guppies. This is the most popular group of pet fishes, and a large percentage of these species are aquacultured.
    2. Temperate freshwater pond fish, such as goldfish and koi (ornamental carp).
    3. Native (US) freshwater species, such as bass, pygmy sunfish, shiners, dace, catfish, and sticklebacks. Certain species in this group may be subject to local fish and game restrictions. Not common as pets in the home.
    4. Tropical marine (saltwater) species, such as clownfish, damselfish, wrasses, butterflyfish, lionfish, marine angelfish, tangs, true sharks, and stingrays. A growing number of species groups are aquacultured, but currently, the majority are wild caught.
    5. Native (US) marine species, such as sculpins, flounder, drum, and herring. Fish in this group are uncommonly kept in the home aquarium and may be subject to local fish and game restrictions.

POPULAR FRESHWATER ORNAMENTAL FISHES

- Some good freshwater community aquarium species: mollies, guppies, swordtails, gouramis, silver dollars, tetras, *Corydoras* catfish, small locariid (plecostomus family) species, danios, and loaches.
- Pond fish: koi and goldfish (there are numerous varieties available).

DOI: 10.1201/9781003057727-1

- Although wild specimens may fare better with water quality parameters similar to their origins, most commonly kept freshwater species are aquacultured and acclimate readily.
- More aggressive aquarium species:
    1. Medium to larger cichlids including Central and South American species, such as oscars, Jack Dempseys, red devil, jewel, green terror, and convicts as well as African cichlids, are often kept in larger, cichlid-only tanks where species, social structure, numbers, and habitat structure must be carefully balanced. For example, more fish (eight or more) are better for African cichlid tanks to help spread/mitigate aggression.
    2. Large *plecostomus* species, piranhas (illegal in some states), and carnivorous catfish.
- Single pet aquarium species: goldfish, any of the aggressive species listed above, pacu, large catfish.

The Tetras: Family Characidae (Most); Also Serrasalmidae and Gastropelecidae (Figures 1.1–1.5)

Figure 1.1 A neon tetra (*Paracheirodon innesi*). Courtesy FTFFA

Figure 1.2 A cardinal tetra (*Paracheirodon axelrodi*). Courtesy FTFFA

Figure 1.3 Several serpae tetra (*Hyphessobrycon eques*). Courtesy FTFFA

Figure 1.4 Two emperor tetra (*Nematobrycon palmeri*). Courtesy FTFFA

Figure 1.5 Two silver dollars (*Metynnis* sp.). Courtesy 5-D Tropical Inc.

- Several hundred different species; bright colors (many silver, black, red, blue).
- Many are small fish (1–2 inches [2.5–5 cm]). Have swim bladders and adipose fins (small fleshy fin behind the dorsal fin). Some species (pacu, piranha, silver dollars) can get much larger (5–35 inches [2–14 cm]).
- Active swimmers; most smaller species coexist well with other fish.
- Egg layers; some may be difficult to breed.

Common tetra species

Black neon tetra	*Hyphessobrycon herbertaxelrodi*
Black phantom tetra	*Hyphessobrycon megalopterus*
Black tetra	*Gymnocorymbus ternetzi*
Bleeding heart tetra	*Hyphessobrycon erythrostigma*
Blind cave tetra	*Astyanax fasciatus mexicanus*
Blood fin tetra	*Aphyocharax anisitsi*
Buenos Aires tetra	*Psalidodon anisitsi*
Cardinal tetra	*Paracheirodon axelrodi*
Glowlight tetra	*Hemigrammus erythrozonus*
Common hatchet	*Gasteropelecus sternicla*
Lemon tetra	*Hyphessobrycon pulchripinnis*
Neon tetra	*Paracheirodon innesi*
Serpae tetra	*Hyphessobrycon eques*
Red pacu	*Piaractus brachypomum*
Silver dollar	*Metynnis argenteus*
Red-bellied piranha	*Pygocentrus nattereri*
Rummy-nosed tetra	*Hemigrammus bleheri*

The Carps and Barbs: Family Cyprinidae (Figures 1.6–1.12)

- Many similar to tetras in size; less colorful (silver, gold, black/red).
- Some have barbs (small fleshy whiskers on lower jaw). Most lack an adipose fin. Have swim bladders.
- Active swimmers.
- Egg layers (in fine-leaved plants); many easy to breed.
- Warmwater: barbs, danios, freshwater sharks, and rasboras.

Figure 1.6 Long-finned zebra danio (*Danio rerio*). Courtesy FTFFA

Figure 1.7 Rosy barb (*Pethia conchonius*). Courtesy FTFFA

Figure 1.8 Rainbow shark (*Epalzeorhynchos frenatum*). Courtesy FTFFA

Figure 1.9 Redtail black shark (*Epalzeorhynchos bicolor*). Courtesy FTFFA

Figure 1.10 School of comet goldfish (*Carassius auratus*). Courtesy 5-D Tropical Inc.

Figure 1.11   Redcap oranda goldfish (*Carassius auratus*). Courtesy 5-D Tropical Inc.

Figure 1.12   Juvenile koi (*Cyprinus rubrofuscus*). Courtesy 5-D Tropical Inc.

- Temperate: goldfish and koi are among the more commonly seen aquarium fish in private practice.
- Goldfish:
    - More than 100 varieties.
    - Get much larger than most common tropical fish; don't keep with warmer water species.
    - Males often have raised tubercles on the gill plate and leading rays of the pectoral fins during breeding season. Spring breeders. Chasing is not reliable in determining sex. Females spawn in plants. Chasing precedes spawning (cover rough edges of pond decorations/substrate to reduce risk of injury).
- Koi:
    - Strong, hardy, tolerate temperature variation, if not severe.
    - Grow to fit size of their habitat (up to 30 inches [76 cm] long).
    - At 1 inch (2.54 cm) long, most are black or gray; change color as they grow, but general pattern is established when they reach 5 inches (12.7 cm) in length.

Common carps and barb species

Goldfish	*Carassius auratus*
Koi	*Cyprinus carpio/Cyprinus rubrofuscus*
Clown barb	*Puntius everetti*
Zebra danio	*Danio rerio*
Rosy barb	*Pethia conchonius*
Harlequin rasbora	*Trigonostigma heteromorpha*
Tiger barb	*Puntigrus tetrazona*
Cherry barb	*Puntius titteya*
Tinfoil barb	*Barbonymus schwanenfeldii*
Flying fox	*Epalzeorhynchos kalopterus*
Rainbow shark	*Epalzeorhynchos frenatum*
Red-tailed black shark	*Epalzeorhynchos bicolor*
Silver/Bala shark	*Balantiocheilos melanopterus*

The Catfishes (Figures 1.13–1.15)

Figure 1.13   Common pleco (*Hypostomus plecostomus*). Courtesy FTFFA

Figure 1.14   Peppered cory cat (*Corydoras paleatus*). Courtesy FTFFA

Figure 1.15 Albino bristlenose catfish (*Ancistrus temminckii*). Courtesy FTFFA

- Order Siluriformes, comprised of 39 different catfish families, mostly freshwater; more than 2,000 different species.
- Some more common families seen in the hobby: Loricariidae (plecos, *Otocinclus*), Callichthyidae (*Corydoras*), Mochokidae (upside-down catfish), Pimelodidae (angelicus catfish). Other families also represented but with fewer species.
- Colors: brown, black, silver, and white.
- All have barbels (whiskers). Have swim bladder. All are scaleless: some are "naked" with no external bony scales or other structures; some have bony/armored plates.
- Size: 1 inch to 5 feet (2.5 cm to 1.5 m), although most common species in the aquarium hobby are 2–6 inches (5–15 cm).
- Bottom fish – actively search for food using their whiskers. Some are nocturnal. Many catfish are piscivorous (fish eaters). If the mouth is large and/or points forward, it is predatory (e.g., angelicus catfish can eat fish larger than itself).
- Some species (e.g., *Corydoras* spp.) will swim to the surface for a gulp of air which they can process through secondary respiratory organs (e.g., the gastrointestinal tract).
- Often sold as scavengers for community tanks to eat left over food and detritus, but for a complete diet, many require more catfish specific diet (use sinking food).

Common catfish species

Aeneus/Bronze cory	*Corydoras aeneus*
Arcuatus/Skunk/Arched cory	*Corydoras arcuatus*
Bandit cory	*Corydoras metae*

Banjo catfish	*Bunocephalus coracoideus*
Blue-eyed plecostomus	*Panaque suttoni*
Chaca chaca catfish	*Chaca bankanensis*
Glass catfish	*Kryptopterus bicirrhis*
Julii cory	*Corydoras julii*
Otocinclus	*Otocinclus vittatus*
Peppered cory	*Corydoras paleatus*
Pictus/Angelicus catfish	*Pimelodus pictus*
Pleco/Plecostomus	*Hypostomus plecostomus/Pterygoplichthys* sp.
Red-tailed catfish	*Phractocephalus hemioliopterus*
Spotted talking/Raphael catfish	*Agamyxis pectinifrons*
Striped talking/Raphael catfish	*Platydoras costatus*
Royal pleco	*Panaque nigrolineatus*
Upside-down catfish	*Synodontis nigriventris*
Zebra pleco	*Hypancistrus zebra*

The Loaches: Families Cobitidae, Botiidae, and Gyrinocheilidae (Figures 1.16–1.18)

- Similar to catfish in ecology but longer and more rounded; some have worm or eel shape (e.g., kuhli loach). Swim bladders present.
- Same colors as the catfish, with some brightly colored.
- Many are secretive, coming out only to feed at night. May show unusual behaviors (e.g., clown loaches may sleep or rest on their side and appear dead).

Figure 1.16 Clown loach (*Chromobotia macracanthus*). Courtesy FTFFA

CHAPTER 1: AQUARIUM FISH FAMILIES AND SPECIES

Figure 1.17   Two zebra loaches (*Botia striata*). Courtesy 5-D Tropical Inc.

Figure 1.18   Pakistani loach (*Botia almorhae/lohachata*). Courtesy 5-D Tropical Inc.

Common loaches species

Algae eater loach	*Gyrinocheilus aymonieri*
Clown loach	*Chromobotia macracanthus*
Dojo loach	*Misgurnus anguillicaudatus*
Kuhli/Coolie loach	*Acanthophthalmus kuhlii*
Pakistani loach	*Botia almorhae/lohachata*
Red-tail botia or Orange-finned loach	*Yasuhikotakia modesta*
Zebra loach	*Botia striata*

The Livebearers: Family Poeciliidae (Figures 1.19–1.22)

- Many live/can live in water with some salt. Some species (e.g., sailfin molly) may do better with a small amount of sea salt added to their water.
- Small to medium size (~1–5 inches [2.5–12.5 cm]); the male is often smaller and more colorful than the female.

POPULAR FRESHWATER ORNAMENTAL FISHES

Figure 1.19   A male guppy (*Poecilia reticulata*). Courtesy FTFFA

Figure 1.20   A male swordtail (*Xiphophorus hellerii*). Courtesy FTFFA

Figure 1.21   Gold dust molly (*Poecilia* sp.). Courtesy FTFFA

Figure 1.22   Male sunburst platy (*Xiphophorus* sp.). Courtesy FTFFA

- Wide variety of color strains with different kinds of finnage and body shape.
- Many have an enlarged dorsal fin. Young fish and females have fan-shaped anal fins; at about 2 months, the male anal fin becomes narrow and tube shaped, becoming the gonopodium, which is used to transfer sperm packets to the female. Swim bladders present.
- Give birth to live young.
- Will reproduce in a community tank, but parents and other fish may eat fry. Females may hold sperm packets for weeks to months.
- Carnivorous, feeding on insect larvae in the wild (used in mosquito control programs). Exception: mollies need vegetable matter. Can feed flake food.

Common livebearer species

Guppy/Millions fish	*Poecilia reticulata*
Southern platyfish	*Xiphophorus maculatus*
Swordtail	*Xiphophorus hellerii*
Variable platy	*Xiphophorus variatus*
Sailfin molly	*Poecilia latipinna*

The Cichlids: Family Cichlidae (Figures 1.23–1.27)

- Includes fish with special water requirements; less tolerant of laxity in water quality than other tropicals.
- Size: most are 1–10 inches (2.5–25.5 cm). Many have a classical "fish shape" with prominent fins, but species can be very diverse in shape, finnage, and coloration.

Figure 1.23   Peacock cichlid (*Aulonocara* sp.). Courtesy FTFFA

Figure 1.24 Lake Tanganyika cichlid (*Neolamprologus brichardi*). Courtesy FTFFA

Figure 1.25 Freshwater angelfish (*Pterophyllum scalare*). Courtesy FTFFA

Figure 1.26 Blue ram dwarf cichlid (*Mikrogeophagus ramirezi*). Courtesy FTFFA

Figure 1.27   Oscar cichlid (*Astronotus ocellatus*). Courtesy FTFFA

- Prized species include the more rounded discus and the triangular freshwater angelfish.
- Many colors.
- Vary in their reproductive methods. Have swim bladders. Very intelligent.
- Most species, e.g., African Rift Lake cichlids, can be very territorial and aggressive and may benefit from higher numbers/densities to help reduce and diffuse social aggression.

Common major cichlid groups include:
- Old World – most common are African Rift Lake cichlids. Rift Lake cichlids include the brightly colored species often kept in mixed groups. Typically prefer high hardness and high pH. Many in this group are mouthbrooders, holding eggs and young in their mouths for protection.
- New World – Central, South American, and North American cichlids. This group includes the larger Central and South American cichlids including the popular "oscar," green severum, and Jack Dempsey, many of which come from more moderate levels of pH and hardness. This group also includes angelfish and discus species, which come from waters with lower pH and hardness. Primarily substrate spawners.
  - Angelfish:
    - Schooling fish.
    - Grow to 9 inches tall (23 cm), need large (>20 gallon; 75 L) aquarium.
    - Other species, such as barbs, may attack angels.
    - Angels will eat smaller fish and their own fry.
  - Discus:
    - Prefer neutral to more acidic pH and warmer temperature (82–85°F).
    - Very sensitive to poor water quality conditions.
- Dwarf cichlids (from South America and Africa), including the blue ram and rainbow krib, are, as adults, on the smaller end of the size range.

Dwarf South American cichlids come from waters with lower pH and hardness, whereas dwarf African cichlids may be more varied in water quality preferences. Primarily substrate spawners.

Common cichlid species
Old world cichlids

Jewel cichlid	*Hemichromis bimaculatus*
Kennyi	*Pseudotropheus lombardoi*
Malawi eye-biter	*Dimidiochromis compressiceps*
Peacock cichlids	*Aulonocara* spp.
Powder blue/pindani	*Pseudotropheus socolofi*
Zebra cichlid	*Pseudotropheus zebra*

New world cichlids

Angelfish	*Pterophyllum scalare*
Black-belt cichlid	*Vieja maculicauda*
Blue acara	*Aequidens pulcher*
Convict cichlid	*Amatitlania nigrofasciata*
Discus	*Symphysodon* spp.
Eartheater/Jurupari cichlid	*Satanoperca jurupari*
Firemouth	*Thorichthys meeki*
Green terror	*Andinoacara rivulatus*
Jack Dempsey	*Rocio octofasciata*
Oscar	*Astronotus ocellatus*
Pike cichlid	*Crenicichla* sp.
Rainbow cichlid	*Herotilapia multispinosa*
Red devil	*Amphilophus labiatus*
Severum	*Heros severus*

Dwarf cichlids

Cockatoo cichlid	*Apistogramma cacatuoides*
Ram dwarf cichlid	*Mikrogeophagus ramirezi*
Krib/Kribensis	*Pelvicachromis pulcher*

The Labyrinth Fish: Family Osphronemidae and Helostomatidae (Figures 1.28–1.29)

- All possess a labyrinth organ or pseudo-lung, found in the head, which facilitates the breathing of atmospheric air.
- Size: generally 2–4 inches (5–10 cm), laterally compressed.
- Swim bladders present.
- Brightly colored (some females are less colorful than males).
- Many common male gourami species have a longer, more pointed dorsal fin (females' are more rounded).

Figure 1.28   Dwarf gourami (*Trichogaster lalius*). Courtesy FTFFA

Figure 1.29   Blue paradise fish (*Macropodus opercularis*). Courtesy FTFFA

- Prefer the top portion of the tank; require floating food. Enjoy floating plants.
- Males build nest at the surface of the water using bubbles of air and saliva. Fish spawn under the nest, then male places the eggs in the nest, guarding and caring for them (and sometimes the new fry) while they develop. Male becomes very aggressive when guarding and will even attack the female spawning partner.

Common labyrinth fish species

Blue/Three-spot gourami	*Trichopodus trichopterus*
Croaking gourami	*Trichopsis vittata*
Dwarf gourami	*Trichogaster lalius*
Giant gourami	*Osphronemus goramy*
Kissing gourami	*Helostoma temminkii*
Moonlight gourami	*Trichopodus microlepis*
Paradise fish	*Macropodus opercularis*
Pearl gourami	*Trichopodus leerii*
Siamese fighting fish/betta	*Betta splendens*

The Rainbowfish: Family Melanotaeniidae (Figures 1.30–1.31)

- Native to Australia, New Guinea, and Indonesia.
- Males very brightly colored, especially when sexually mature.
- Great for planted tanks and good dither fish for cichlid tanks.

Figure 1.30   Red rainbowfish (*Glossolepis incisus*). Courtesy FTFFA

Figure 1.31   Turquoise rainbowfish (*Melanotaenia lacustris*). Courtesy FTFFA

Common rainbowfish species

Boesemani rainbow	*Melanotaenia boesmani*
Eastern rainbow	*Melanotaenia splendida*
Forktail blue eye rainbow	*Pseudomugil furcatus*
Red rainbow	*Glossolepis incisus*
Turquoise rainbow	*Melanotaenia lacustris*

POPULAR POND FISHES

The Carps: Family Cyprinidae (Described Earlier)

- Koi (Nishikigoi).
- Carp.
- Goldfish.
- Tench.

POPULAR TROPICAL MARINE ORNAMENTAL FISHES

In general, marine tropical fish are considered more difficult to keep than common freshwater species. Strict attention to water quality, social and environmental structure, and dietary needs are critical. The majority of marine tropical fish in human care are wild caught. Marine aquarium fish have proven more difficult to breed in human care due to challenges with broodstock nutrition, early life history traits (many have a prolonged "pelagic"/"planktonic" stage) requiring difficult to raise zooplankton as diets, and unknown maturation cues.

However, an increasing number are aquacultured and commercially available because of the ease of breeding species with monogamous pairs, demersal spawning (eggs laid on a substrate vs. pelagic – into the water), precocial larvae ("more mature at hatch"), and extended parental care. Commercially available, cultured marine tropicals include clownfishes (which have been in commercial production since the late 1980s), dottybacks, gobies, blennies, cardinalfishes, and seahorses. Some species of angelfishes, dragonets, filefishes, tangs, small sharks, and a few other species from other families have also begun to enter the commercial market.

Because most of the commonly kept marine fish are hermaphroditic, conditions (environmental, social) may result in sex change of a given individual.

Another consideration is territoriality. A number of species are territorial and semi-aggressive to very aggressive, so species compatibility and groupings, final fish adult size, tank size, and habitat complexity (hiding places, structures) should be taken into consideration. Fish that are more territorial should be added to the tank last.

The Damselfishes and Clownfishes: Family Pomacentridae (Figures 1.32–1.34)

- In general, hardy, colorful, easy to care for, and small.
- Damselfish:
  - Most are delicate and better kept in schools (e.g., yellowtail blue damselfish).
  - Not as easily aquacultured as clownfish species.
- Clownfish:
  - Protandrous hermaphrodites (male first), with one in a group/pair developing into a female (becoming largest).
  - Many species aquacultured.

Figure 1.32   False clownfish (*Amphiprion ocellaris*). Courtesy ORA

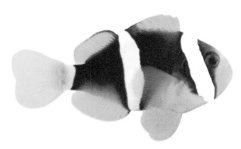

Figure 1.33 Clark's clownfish (*Amphiprion clarkii*). Courtesy ORA

Figure 1.34 Yellowtail damselfish (*Chrysiptera parasema*). Courtesy ORA

- Numerous manmade color and pattern variants.
- > 20 gallons (76L) recommended for tank size.
- Larger species are hardy. Up to 4 inches (10 cm) (e.g., tomato, fire, and sebae clown) in size.
- Do not require an anemone to live in.
- Not usually territorially aggressive unless mated pair.
- False (ocellaris) clownfish may be a bit hardier than true (percula).

Common clownfish and damselfish species

Bicolor damselfish	*Stegastes partitus*
Blue devil	*Chrysiptera cyanea*
Yellowtail damselfish	*Chrysiptera parasema*
Clark's clownfish	*Amphiprion clarkii*
Domino/Three-spot damselfish	*Dascyllus trimaculatus*
False clownfish	*Amphiprion ocellaris*

Fire clownfish	*Amphiprion ephippium*
Green chromis	*Chromis viridis*
Maroon clownfish	*Premnas biaculeatus*
Percula/True clownfish	*Amphiprion percula*
Pink (skunk) clownfish	*Amphiprion perideraion*
Sebae clownfish	*Amphiprion sebae*
Skunk clownfish	*Amphiprion perideraion*
Tomato clownfish	*Amphiprion frenatus*
Yellow-tailed blue damselfish	*Chrysiptera hemicyaneus*

The Dottybacks: Family Pseudochromidae (Figures 1.35–1.36)

- Smaller and more elongated than the pomacentrids; bright coloration.
- Males often larger.
- Protogynous hermaphrodites (female first, later develop into male).
- Can be semi-aggressive and territorial.
- Some species aquacultured (orchid dottyback, neon dottyback, and splendid dottyback).

Figure 1.35   Neon dottyback (*Pseudochromis aldabraensis*). Courtesy ORA

Figure 1.36   Yellow dottyback (*Pseudochromis fuscus*). Courtesy ORA

Common dottyback species

Orchid dottyback	*Pseudochromis fridmani*
Neon dottyback	*Pseudochromis aldabraensis*
Splendid dottyback	*Manonichthys splendens*
Yellow dottyback	*Pseudochromis fuscus*

The Cardinalfish: Family Apogonidae (Figures 1.37–1.38)

- Hardy; grow to 3–4 inches (7.5 cm).
- Nocturnal, with large eyes; require hiding places, but can acclimate to day feeding.
  - Turn off lights when introducing fish to tank to avoid shock.

Figure 1.37   Banggai cardinalfish (*Pterapogon kauderni*). Courtesy ORA

Figure 1.38   Pajama cardinalfish (*Sphaeramia nematoptera*). Courtesy ORA

- Some species may become aggressive to conspecifics, but some, e.g., Banggai cardinalfish, especially as juveniles, do well in groups.
- Many are male mouthbrooders and have large mouths.
- Some species aquacultured.

Common cardinalfish species

Banggai cardinal	*Pterapogon kauderni*
Flamefish cardinal	*Apogon maculatus*
(Gold) Striped cardinal	*Apogon cyanosoma*
Pajama cardinal	*Sphaeramia nematoptera*
Red cardinal (15 spp.)	*Apogon quadrisquamatus*

The Blennies: Family Blenniidae (Figures 1.39–1.40)

- Most lack a swim bladder and swim at the tank bottom.
- Have a long dorsal fin and cirri (hair-like filaments) above the eyes.
- Fang blennies (including *Meiacanthus* spp.) have a swim bladder, swim mid-column, and many, but not all, have a buccal venom gland.
- Hardy; grow to 4 inches (10 cm). Large "canine" teeth.
- Good personality; not as colorful as many gobies; can be territorial and somewhat aggressive.
- Some species aquacultured.

Figure 1.39 Striped blenny (*Meiacanthus grammistes*). Courtesy ORA

Figure 1.40 Kamohara blenny (*Meiacanthus kamoharai*). Courtesy ORA

Common blenny species and variations

Bicolor blenny	*Ecsenius bicolor*
Forktail blenny	*Meiacanthus atrodorsalis*
Golden/Lyretail blenny	*Ecseniusmidas*
Kamohara blenny	*Meiacanthus kamoharai*
Lawnmower blenny	*Salarias fasciatus*
Mimic blenny	*Ecsenius gravieri*
Molly miller blenny	*Scartella cristata*
Redlip blenny	*Ophioblennius atlanticus*
Rockskipper blenny	*Istiblennius zebra*
Striped blenny	*Meiacanthus grammistes*

The Hawkfish (Rockhoppers): Family Cirrhitidae (Figure 1.41)

- Bottom swimmers – no swim bladder. Normally live in deep water; difficult to acclimate to the confines of an aquarium.
- Can be predatory on smaller fish.
- A lot of personality and color.

Common hawkfish species

Flame hawkfish	*Neocirrhites armatus*
Longnose hawkfish	*Oxycirrhites typus*
Redspotted hawkfish	*Amblycirrhitus pinos*

Figure 1.41 Flame hawkfish (*Neocirrhites armatus*). © Central Garden & Pet Company. Reprinted with permission

Popular Tropical Marine Ornamental Fishes

Figure 1.42 Spotted green puffer (*Dichotomyctere nigroviridis*). © Central Garden & Pet Company. Reprinted with permission

The Puffers: Family Tetraodontidae (Figure 1.42)

- Do not keep with invertebrates.
- Grow to up to 12 inches (30.5 cm) long.
- Slow swimming.
- Can be very "nippy" with other fish species.
- The spotted green pufferfish has been aquacultured.

Common puffer species and variations

Burrfish/Striped puffer	*Chilomycterus schoepfi*
Spotted green puffer	*Dichotomyctere nigroviridis*
Dwarf/Pygmy puffer	*Carinotetraodon travancoricus*
Spotted puffer	*Tetraodon nigrifilis*
Whitespotted puffer	*Arothron hispidus*

The Gobies: Family Gobiidae (Figures 1.43–1.44)

- Many small, up to 3 inches (7.5 cm).
- No swim bladder.
- Can be very colorful.
- In some species, fused pelvic fins form a suction disk.
- Spend a lot of time resting on tank floor.
- Neon gobies are well-known "cleaner fish" spending time picking parasites off other fish species and are one of several gobies aquacultured.

Figure 1.43 Neon goby (*Elacatinus oceanops*). © Central Garden & Pet Company. Reprinted with permission

Figure 1.44 Watchman goby (*Cryptocentrus cinctus*). © Central Garden & Pet Company. Reprinted with permission

Common goby species and variations

(Atlantic) Neon goby	*Elacatinus oceanops*
Bumblebee goby	*Brachygobius xanthozona*
Fire goby	*Nemateleotris magnifica*
Golden-head sleeper goby	*Valenciennea strigata*
Yellow shrimp goby	*Cryptocentrus cinctus*

The Dragonets: Family Callionymidae (Figure 1.45)

- Popular species are small and very colorful.
- Some species called "gobies," but these are not true gobies.
- Benthic – stay on the tank floor.
- Wild-caught specimens can be difficult to train to feed on commercial diets; aquacultured specimens much easier to keep.
- Shy.

Popular Tropical Marine Ornamental Fishes

Figure 1.45 Mandarin goby (*Synchirops splendidus*). © Central Garden & Pet Company. Reprinted with permission

Common dragonet species and variations

Mandarin goby (dragonet)	*Synchirops splendidus*
Spotted mandarin	*Synchirops picturatus*

The Butterflyfishes: Family Chaetodontidae (Figure 1.46)

- Chaeto ("hair, bristle") + odont (tooth); many have more specialized feeding and diets including algae, anemones, sponges, and coral (e.g., some species of *Chaetodon*) and, therefore, are not recommended for reef tank systems; some omnivorous.
- Species size range from 4 to 12 inches.

Figure 1.46 Copperband butterflyfish (*Chelmon rostratus*). © Central Garden & Pet Company. Reprinted with permission

- Schooling bannerfish considered relatively reef safe (but may be confused with similar appearing pennant coralfish which is not), as are species of pyramid butterflyfish.
- Wild-caught fish considered difficult to keep and to feed.
- Similar in appearance to angelfish but without spines.

Common butterflyfish species

Raccoon butterflyfish	*Chaetodon lunula*
Copperband butterflyfish	*Chelmon rostratus*
Yellow pyramid butterflyfish	*Hemitaurichthys polylepis*
Schooling bannerfish	*Heniochus diphreutes*
Pennant coralfish	*Heniochus acuminatus*

The Wrasses: Family Labridae (Figure 1.47)

- Very large family; varied shapes and coloration; many smaller.
- Have protractile mouths.
- Very popular group in the hobby.
- Most are protogynous hermaphrodites (female first) and haremic (several females with one dominant male, +/- younger males) breeding groups.
- Cleaner wrasse known to "clean" other fish of parasites.
- Some species have been aquacultured, but only a few are available commercially.

Figure 1.47   Cleaner wrasse (*Labroides dimidiatus*). © Central Garden & Pet Company. Reprinted with permission

Common wrasse species

Cuban/Spotfin hogfish	*Bodianus pulchellus*
Yellowtail wrasse	*Coris gaimard*
Cleaner wrasse	*Labroides dimidiatus*
Bluehead wrasse	*Thalassoma bifasciatum*
Canary wrasse	*Halichoeres chrysus*
Tailspot wrasse	*Halichoeres melanurus*

The Triggerfish: Family Balistidae (Figure 1.48)

- Require large aquarium; most species 8–20 inches (20–50 cm) in size.
- Anterior dorsal fin comprised of three spines, with the first being the longest. The second serves as a trigger locking the first into place.
- Strong jaw with crushing teeth designed for their crustacean, mollusk, and echinoderm (among others) diet.
- Intelligent, aggressive, and can be very territorial.

Common triggerfish species

Clown trigger	*Balistoides conspicillum*
Huma/Picasso trigger	*Rhinecanthus aculeatus*
Niger/Redtooth trigger	*Odonus niger*
Undulate trigger	*Balistapus undulatus*

Figure 1.48 Clown triggerfish (*Balistoides conspicillum*). © Central Garden & Pet Company. Reprinted with permission

The Groupers and Basslets: Family Serranidae and Family Grammatidae (Figures 1.49–1.51)

Figure 1.49   Panther grouper (*Cromileptes altivelis*). © Central Garden & Pet Company. Reprinted with permission

Figure 1.50   Swalesi basslet (*Liopropoma swalesi*). © Central Garden & Pet Company. Reprinted with permission

Figure 1.51   Royal gramma (*Gramma loreto*). © Central Garden & Pet Company. Reprinted with permission

- True groupers are large and will eat other fish; best kept with triggers, lionfish, and moray eels.
- Basslets, anthias, and grammas are smaller and compatible with most other fish.
- Anthias and grammas are planktivores and considered "reef-tank friendly."
- Both families are protogynous hermaphrodites (female first but can switch to male). This can be problematic for anthias due to potential intraspecies aggression.

Common serranid species

Panther grouper	*Cromileptes altivelis*
Swalesi basslet	*Liopropoma swalesi*
Bartlett's anthias	*Pseudanthias bartlettorum*
Royal gramma	*Gramma loreto*

The Marine Angelfishes: Family Pomacanthidae (Figure 1.52)

- Poma (<cover (i.e., operculum) + acanth (<thorn); marine angelfish have a spine on their opercula.
- Dwarf angelfish (*Centropyge* spp.) grow up to 15 cm; queen angelfish and French angelfish grow up to 46 cm.
- Considered more difficult to keep, and some species are not recommended for reef tanks.
- Angelfish are protogynous hermaphrodites (female first and may transition to male under the right conditions).
- Juveniles have different coloration/phases from adults.
- Many dwarf angelfish (*Centropyge* spp.) can be easier to keep than larger angels, and a few species are aquacultured.
- Larger angelfish require much more space (200–300 gallons or more).
- All species can be very territorial and aggressive to conspecifics and other fish of similar size.
- Pomacanthids are susceptible to lateral line depigmentation (formerly head and lateral line erosion (HLLES).

Figure 1.52　French angelfish (*Pomacanthus paru*). © Central Garden & Pet Company. Reprinted with permission

Common marine angelfish species

Cherub angelfish	*Centropyge argi*
Coral beauty angelfish	*Centropyge bispinosa*
Lemonpeel angelfish	*Centropyge flavissima*
Queen angelfish	*Holacanthus ciliaris*
French angelfish	*Pomacanthus paru*

The Tangs and Surgeonfish: Family Acanthuridae (Figure 1.53–1.55)

- Members of this family have sharp spines (acanth (< thorn)) on their caudal peduncle and tough skin.
- Considered more difficult to keep; species may have more specific dietary preferences/requirements.
- Some species (e.g., yellow tang and Pacific blue tang) have been aquacultured.

Figure 1.53 Pacific blue tang (*Paracanthurus hepatus*). © Central Garden & Pet Company. Reprinted with permission

Figure 1.54 Yellow tang (*Zebrasoma flavescens*). © Central Garden & Pet Company. Reprinted with permission

Figure 1.55 Juvenile aquacultured yellow tang (*Zebrasoma flavescens*). GA Lewbart

- Reef-dwellers grow up to 10 inches (25 cm); very colorful.
- Acanthurids are susceptible to lateral line depigmentation (formerly HLLES).

Common tang and surgeon species and variations

Brown tang	*Acanthurus nigrofuscus*
Goldrimmed surgeon	*Acanthurus nigricans*
Naso tang	*Naso lituratus*
Pacific blue tang	*Paracanthurus hepatus*
Purple surgeon	*Acanthurus xanthopterus*
Purple tang	*Zebrasoma xanthurus*
Sailfin tang	*Zebrasoma veliferum*
Yellow tang	*Zebrasoma flavescens*
Yellow-tailed surgeon	*Prionurus laticlavius*

POPULAR AQUARIUM NON-FISH SPECIES

Freshwater

- Aquarium snails: *Lymnaea* (can be intermediate host for trematodes, which require a bird to complete life cycle in most cases).
- Shrimp (especially *Neocaridina* and *Caridina* spp.)
- Crayfish.

Marine Tanks

- Shrimps (cleaner, candy cane, blood, and coral-banded).
- Crabs (anemone, hermit, and arrow).
- Sea stars.
- Sea urchins.
- Marine snails (*Turbo* spp., *Astraea* spp.).

LAWS AND PERMITS

- Few laws govern the keeping of pet fish.
- Restrictions may apply to protected or prohibited species that may have special conservation concerns or are regulated for other reasons (e.g., piranhas and stingrays in Florida and Texas) or to species listed by the Convention on International Trade in Endangered Species (CITES) (very few). Most laws affect breeders, importers, and commercial aquariums.
- Contact the US Fish and Wildlife Service (www.fws.gov/permits), state wildlife, conservation, or agricultural offices or international wildlife organizations for details.
- In the United States, the Food and Drug Administration (FDA) and Environmental Protection Agency (EPA) regulate the use of drugs and chemicals (including pesticides that may be used for disease treatments).
- Consult the EPA and local authorities for regulations regarding runoff of pond water that has been treated.

CHAPTER 2
AQUARIUM SYSTEMS AND WATER QUALITY

IMPORTANT SYSTEM PROCESSES

Ultimately, the nitrogen cycle as well as feed, dead or dying animals and plants, and other organics contribute to compounds and particulates that are processed in a well-designed filtration system. All recirculating aquaculture systems – whether home or public display aquariums, koi ponds, retail or wholesale aquarium systems, or other aquaculture systems that re-use and filter water – should have the processes (**Figure 2.1a**) and related components described below. These processes should also be examined when water quality issues arise.

- Circulation (pump and pipes).
    - Movement of water into, through, and out of the tanks and filtration. Will vary depending on the pump(s) and pipework/plumbing, number of fish holding units (e.g., single aquarium vs. multi-tank retail system), and filtration components and their configuration, i.e., in series or in parallel, each with advantages and disadvantages (**Figures 2.1b–2.1c**).
- Solids capture ("mechanical filter") (**Figure 2.1d**).
    - Different types of solids (settleable, suspended, and fine) exist.
    - Their removal enhances water clarity and reduces their potential a) to breakdown into harmful compounds, b) to damage external tissues of the fish including gills, and c) to serve as reservoirs or food for pathogens.
    - A variety of different materials, configurations, and sizes can be effective depending on solid type and system requirements. Not all mechanical filters remove all solid types – especially fine – so water changes may help.
    - Small hobbyist aquariums often use a fiber/cloth material in a sheet or pleated cartridge form or a sponge media to trap particulates that flow through them. These then need to be cleaned or changed. Recirculating ponds or larger systems may have a settling basin, or solids may settle in areas of lower flow and require siphoning. Protein skimmers remove dissolved proteins/macromolecules.
- Biofiltration (biofilter) (**Figures 2.1e–2.1g**).

DOI: 10.1201/9781003057727-2

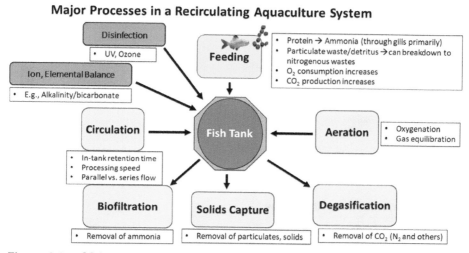

Figure 2.1a   Major processes in a recirculating aquaculture system

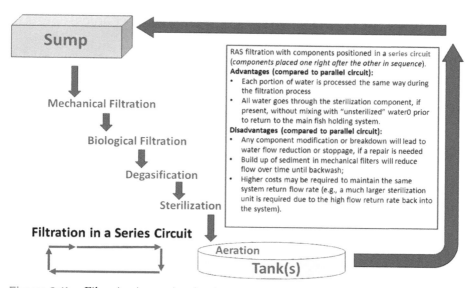

Figure 2.1b   Filtration in a series circuit

- A variety of different microbes that comprise the biofilter community remove harmful ammonia and nitrites (**Figure 2.1e**; more details provided further below in "The Nitrogen Cycle in Aquarium and Pond Systems" section). Some metabolize ammonia to nitrites, some metabolize nitrites to nitrates, some individual species can do both. Many of

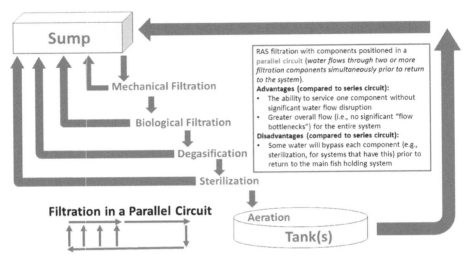

Figure 2.1c Filtration in a parallel circuit

Different Types of Solids and Corresponding Mechanical Filters

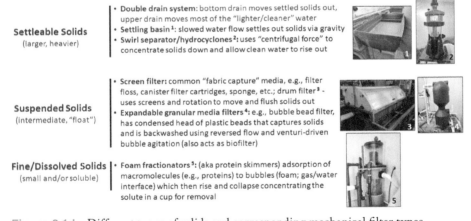

Figure 2.1d Different types of solids and corresponding mechanical filter types

these – the "autotrophic" bacterial species – use bicarbonate (an important and common component of alkalinity) as a carbon source.
- The biofiltration process requires ammonia, oxygen, and bicarbonate and produces carbon dioxide and acid that must be managed. The more common biofilter microbes prefer a pH range of approximately 7.2–8.2 (~7–9 is functional).

Important System Processes

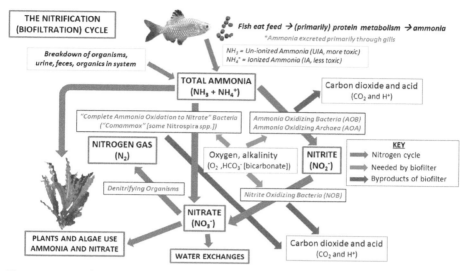

Figure 2.1e The nitrification biofiltration cycle

Biological Filtration: **Thick** vs. Thin Biofilm

- **Thick Film Biofilters**
  - Autotrophs start, then slow down
  - As heterotrophs overrun
    - Need periodic "cleaning"
- **Thin Film Biofilters**
  - Autotrophs start, then stay revved up
  - Limited build up for heterotrophs
    - "Operational" sloughing ("self-clean")

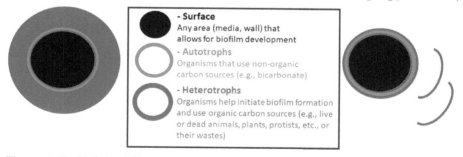

Figure 2.1f Biological filtration – thick vs. thin film

- Commonly used biofilters require a surface and are technically "fixed film" biofilters. Thick film and thin film biofilters are two major categories of fixed film biofilters. Thick film biofilters (which may also act as mechanical filters) require periodic flushing/backwash to help optimize efficiency, while thin film biofilters "self-clean" during their normal operation.

Fixed Film Biofilter Types

Emergent (exposed to air)
- Thin Film ("self-cleaning")
- Rotating Biological Contactor (e.g., biowheel [1])
- Trickle Filter (water trickles through air over media [2])

- Maximizes O_2 transfer
- Media/surface exposed to air
- Relatively low surface area
- Biofilm sloughs from media during normal operation

Submerged (underwater) — Packed
- Thick Film (needs backwashing/cleaning)
- Submerged Rock (e.g., undergravel filter [3])
- Shell Filter or Other Packed Media

- Assumes sufficient DO
- Also acts as mechanical filter
- Needs periodic vacuuming/siphoning

Submerged (underwater) — Expandable
- Thick Film (needs backwashing/cleaning)
- Upflow Sand Filter
- Floating Bead Clarifier/ Bead Filter [4]
- Foam (e.g., sponge filter [5])

- Assumes sufficient DO
- Also acts as mechanical filter
- Media must be temporarily expanded - abrasion or manual compression removes biofilm and solid waste

Expanded
- Thin Film ("self-cleaning")
- Fluidized Sand Bed [6]
- Moving Bed Biofilm Reactor [7] (MBBR)

- Assumes sufficient DO
- Constant working motion with abrasion removes excess biofilm and solids

Figure 2.1g   Fixed film biofilter types

- Aeration (airstones as well as components that provide water flow over/into the tank or facilitate increased exposure of water to air).
  - Increases oxygen levels in the tank.
- Degasification (increased exposure of water to air prior to re-entry into the system; typically includes some water movement through air).
  - Although somewhat related to aeration, this process concentrates on removal of harmful gases that may enter a system during a water change, after biofiltration, in a system with air introduced under pressure or with high organics or unwanted, small anaerobic areas. Gases removed include carbon dioxide, hydrogen sulfide, and excess nitrogen gas.
- Disinfection.
  - Different processes by which water is treated to kill or inactivate specific pathogens. Not common in hobbyist systems but may be in some more advanced systems; used in public aquaria, aquarium fish wholesale/retail, and commercial aquaculture. Ultraviolet sterilization (UV) is more common and safer than ozonation. Both are described briefly below to provide some background; however, if these are components of a system you are evaluating, contact the manufacturer and seek more comprehensive resources and information (e.g., Hadfield and Clayton, 2022).

- Ultraviolet (UV) sterilization units have bulbs that emit a wavelength (~254 nm) that penetrates through the water and organism to target its DNA.
- The effectiveness of a UV sterilizer will depend on a number of factors, including:
  - Zap dose – the "kill" dose required for a specific organism and life stage.
    - Measured in microwatt (μW) sec/cm^2 or milliwatt (mW) sec/cm^2 [or microjoule (μJ)/cm^2 or milliJoule (mJ)/ cm^2].
    - Can be affected by change in water flow.
  - The life cycle of the pathogen and whether adequate numbers of the pathogen will go through the UV unit to have an effect.
  - Turbidity or color of the water (affects UV penetration).
  - Proper management, bulb life, and bulb sleeves (may develop biofilm or precipitation, which also affect UV penetration).
- UV systems are also used by hobbyists to control microalgae (phytoplankton) in koi ponds.
- Ozonation uses ozone (O_3), a highly reactive gas, which forms free radical compounds that disrupt and kill microbes. Ozone, toxic to fish (in water) and to people (in air), is applied in a separate reaction chamber (NOT in the main system) and removed by UV, carbon, off-gassing, or other means prior to return of the water to the main system.
- System water must be monitored (e.g., measure oxidation/reduction potential (ORP) with a meter and compare to an acceptable system-specific standard or use a diethyl-p-phenylene diamine (DPD) colorimetric test) to ensure safe levels. Ozone should also be measured in surrounding air for human health. Ozone is rarely used in hobbyist or retail/wholesale outlets but is common in public aquaria and some aquaculture facilities.
- Ozone in seawater is known to form unwanted products including hypobromous acid, which is toxic, and also highly reactive iodate, a form of iodine, which is not biologically available to fish and may result in iodine deficiency.
- Ozone also has been used to reduce or eliminate color, turbidity, algae, odor, and taste.
- For freshwater and saltwater water chemistry reference ranges, see **Table 2.1**.

Table 2.1 **Water chemistry reference ranges**

	FRESHWATER	SALTWATER
Oxygen (mg/L or ppm)	5–8 mg/L	5–8 mg/L
pH	6.5–8.5	8.0–8.5
Alkalinity	75–200 mg/L	200 mg/L (100–300 mg/L)
Carbon dioxide	<10 mg/L *	<10 mg/L*
Salinity** (g/L or ppt)	0	30–35
Specific gravity	1.0	1.023–1.026
Un-ionized ammonia (UIA)**	<0.05 mg/L	<0.05 mg/L
Nitrite	<0.1 mg/L	N/A****
Nitrate	<30 mg/L	<20 mg/L
Calcium	Varies depending on species, but 50–200 mg/L	400–500 mg/L

*Carbon dioxide levels of 10–12 mg/L may be tolerated by fishes if oxygen levels are high, but species and age sensitivities may exist. In marine systems, free carbon dioxide levels are zero if pH is above 8.34. As pH falls below 8, carbon dioxide concentrations may rise and be problematic for marine fishes that are less likely to be exposed to free carbon dioxide.

** Species' tolerances differ, and some freshwater species may fare well with low levels of salt (e.g., 1–2 g/L) for extended periods; other species are more brackish and thrive at 2–5 g/L or more (e.g., well-known brackish water species include *Monodactylus* spp., spotted green puffer, *Scatophagus* spp.), and these may benefit from use of sea salt vs. sodium chloride.

***Un-ionized ammonia (UIA) is calculated from total ammonia based on temperature and pH values.

****Detectible nitrite levels indicate a problem with the filter. Nitrite is rarely toxic to marine fish except at very high levels because chloride competes with nitrite for uptake at the gill epithelium; however, species and individual sensitivities may exist.

TANK ESSENTIALS

- Size: small tanks (<30 gallons; 115 L) can be more difficult to manage because water quality parameters can change faster. However, larger tanks will require increased filtration capacity and stronger structural support. One gallon (3.78L) of freshwater weighs ~ 8.3 pounds (3.8kg). One gallon (3.78L) of saltwater weighs ~ 8.6 pounds (3.9kg).
- Light: artificial light is easier to control than sunlight (so avoid placing tanks directly by windows), and for many freshwater aquarium systems and fish only marine aquaria, a full spectrum aquarium light should be adequate. However, light meters that measure photosynthetically active radiation (PAR) are useful to determine level of relevant light available for

photosynthesizing organisms (e.g., plants, coral) and specialized aquarium lights are available.
- PAR (measured in µmol/m²/sec) recommendations for freshwater planted tanks: 10–30 easy plants; 30–60 medium; 60 or more advanced plants.
- Do not use only blue or white fluorescent tubes ("warm"); the red spectrum is needed as well to bring out fish color and enhance plant growth. Full-spectrum light (including red and blue ranges) are ideal for planted tanks. Incandescent bulbs are not recommended because they are inefficient sources of light and give off heat.
- Cover: many fish will jump or may easily slip through loose covers, so a tight-fitting hood or cover is essential.
- Filter: used to remove ammonia (nitrogenous waste), suspended debris, and some toxins. Water conditioners remove chlorine but not ammonia. Some conditioners bind ammonia. Larger tanks and tanks with heavier fish loads require greater biological filtration capacity.
- Filter types.
  - Biofiltration/Nitrification (**Figures 2.2–2.7**): hang-on-back filter, undergravel, biowheel, sponge, or wet/dry (trickle) biological filter for larger volumes. Allow several weeks for biological filter development before adding a full complement of fish. (See "Nitrogen Cycle in Fish Systems" section).
  - Mechanical filtration: floss, sponge, canister, other media. Low flow areas (e.g., in ponds) may serve to "settle" out particulates for removal later. Size of particulates removed depends on filter type. Will not remove ammonia or nitrites.

Figure 2.2   Hang-on aquarium back power filter. GLOFISH® images owned by GloFish, LLC and used under license.

Figure 2.3 Hang-on aquarium back power filter with biowheel. GLOFISH® images owned by GloFish, LLC and used under license.

Sponge Aquarium Filter

Figure 2.4 Sponge filter. Depositphotos

Figure 2.5A Sponge filter in African cichlid tank. GA Lewbart

Figure 2.5B Bubble bead filter. RP Yanong

Figure 2.6 Bioballs in wet/dry filter. Brandon Ray

Figure 2.7 Cannister power filter (cutaway view). GLOFISH® images owned by GloFish, LLC and used under license.

- Diatomaceous earth (DE) filter (**Figure 2.8**): can be used as a temporary filter to remove particulates as small as 3-5 micron; will clog more rapidly if used long term.
- Adsorption: activated carbon (**Figure 2.9**). Can be used to help remove dyes, tannins, some toxins, and medications but may not be desired for long term use, as some forms have been linked to lateral line depigmentation. Has very limited capacity. Remove during medical treatment of water.

Figure 2.8 Diatom filter being used. GA Lewbart

Figure 2.9 Activated carbon. GLOFISH® images owned by GloFish, LLC and used under license.

- Protein skimmer/foam fractionator (**Figure 2.10**): unit which uses fine bubbles to remove organic compounds including some nutrients and waste particles from water before they breakdown further into ammonia or other toxic compounds.
- Ultraviolet (UV) (**Figure 2.11**): specially designed filter units that utilize UV bulbs to reduce loads of specific size/type organisms in water based on their required "zap or kill dose"; used in some systems to kill pathogens (viruses, bacteria, some parasites) or microscopic algae.
- Other tank supplies.
  - Gravel: enough to form a 2–3-inch (5–8 cm) layer over undergravel filter (if applicable). Should be vacuumed/cleaned regularly if part of an undergravel filter.
  - Other substrate: if a planted tank, aquarium safe dirt and gravel should be purchased to promote healthy plant growth.

Figure 2.10   Foam fractionator (protein skimmer). RP Yanong

Figure 2.11   Ultraviolet sterilizer filters. GA Lewbart

- Submersible heater (**Figure 2.12**): maintain ideal water temperature (monitor with thermometer). Cold fish will huddle at bottom of tank; warm fish may swim at the surface for better access to oxygen.
- Recommended heater size: 4 watts/gallon (15 watts/L); however, temperature differential between the air temp and desired water temperature

Figure 2.12 Submersible heater. GLOFISH® images owned by GloFish, LLC and used under license.

may necessitate a stronger heater. Broken or cracked heaters may cause the release of dangerous stray voltage.
- Marine tropicals: 78–84°F (25–29°C).
- Freshwater tropicals: 75–80°F (24–26°C).
- Discus: 80–86°F (26–30°C).
- Guppies: 68–84°F (20–29°C).
- Goldfish and koi thrive at room temp (68–72°F [20–22°C]) but can live in outdoor ponds year-round (even when pond freezes over, as long as there is running water and an opening in the ice).
  – Air pump (with tubing, air stones) (**Figures 2.13–2.14**): maintain dissolved oxygen at 5–10 ppm and used to supplement other components if necessary. Oxygen saturation (the maximum amount of oxygen a body of water can hold without any organisms) will depend on temperature, salinity, and pressure (altitude).
- Decrease in dissolved oxygen results from salinity, decreased atmospheric pressure, and increased temperature (cool-water fishes require more

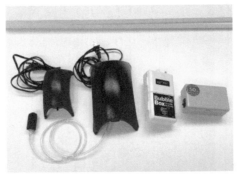

Figure 2.13 Electric and battery operated air pumps. RP Yanong

Figure 2.14 Tubing and air stones. RP Yanong

oxygen than warm-water fishes). Hypoxia develops in heavily planted, shallow, outdoor ponds and warm weather. Test immediately prior to dawn when oxygen levels are lowest.
- Plants: provide a place for fish to hide. Plastic plants are usually not eaten by fish and do not transmit disease (unless transferred from another tank). Live plants can carry disease (acting as fomites for pathogens) and don't do well with undergravel filters or some species of fish.
- Tank maintenance equipment (e.g., bucket, siphon hose, sponge) (**Figures 2.15, 2.16**).
- Cleaning and disinfection: when cleaning out tanks or equipment that will be used in fish tanks, especially after an infectious disease outbreak, surface organics and biofilms should be scrubbed and removed completely prior to use of disinfectants to increase exposure. After disinfectant use, detoxify and rinse as thoroughly as possible. Many common pathogens can be killed with these suggested disinfectants

Figure 2.15   Various sized plastic buckets. RP Yanong

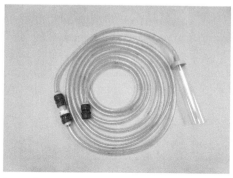

Figure 2.16   Siphon hose. RP Yanong

Figure 2.17 Nets of various sizes. RP Yanong

and concentrations (but some more refractory pathogens, e.g., spore-forming organisms, mycobacteria, and others may require higher doses and durations) and should be rinsed or neutralized thoroughly after use.
- Chlorine/bleach (sodium hypochlorite – liquid or calcium hypochlorite – granular): 200–500 mg/L available chlorine, 10–60 min; not recommended for nets or other delicate equipment; use products with no perfume or other additives; neutralize with sodium thiosulfate.
- Virkon®: 0.5%–2%, for 10–15 min (use as per manufacturer's recommendations); can be used as a net dip; rinse after disinfection, prior to use.
- Hydrogen peroxide: 3%–5% for 5–15 min; may be used as a net dip, rinse thoroughly after disinfection, prior to use.
- Quaternary ammonium compounds (QACs including benzalkonium chlorides, e.g., Roccal®-D): 250–500 mg/L for 10–30 min; rinse thoroughly after use; may be used as a net dip, but not effective against pseudomonads, and nets must be rinsed thoroughly after disinfection, prior to use.
- Alcohol (ethyl or isopropyl): 60%–90% (v/v) for 10–30 min.
- For additional information, see Yanong and Ehrlacher-Reid, 2012.
• Nets (**Figure 2.17**): different sizes, mesh types, and other types of nets will be necessary based on fish size, external features (e.g., spines vs. smooth bodied). Special, softer, and longer "koi nets" may also be available and useful for pond calls.

THE NITROGEN CYCLE IN AQUARIUM AND POND SYSTEMS

Although most of the nitrogen cycling (described in **Figure 2.1e**) occurs in the biofilter unit and/or with the help of plants or algae, this process can occur on any biofilm surfaces (e.g., on the walls of the tank, within pipework, etc.) with access to ammonia, bicarbonate, and oxygen.

Ammonia, the breakdown product primarily from protein metabolism, is excreted primarily from the gills of most teleost fishes. In systems with adequate biofiltration, this ammonia, which is toxic, is metabolized by several groups of bacteria, with the autotrophic bacteria being the most beneficial (see **Figure 2.1e**).

1. First, ammonia oxidizing bacteria (AOB) and ammonia oxidizing archaea (AOA), which require oxygen and bicarbonate (as a carbon source), metabolize ammonia to nitrite. Nitrite is also toxic.
2. Then, nitrite oxidizing bacteria (NOB), which also require oxygen and bicarbonate, metabolize nitrite to nitrate, considered much less toxic.
3. Recently, some *Nitrospira* sp. bacteria with the ability to completely oxidize ammonia to nitrate have been identified (Comammox).
4. Other anaerobic microorganisms, if present under certain conditions, can also play a role in ammonia and nitrite removal.
5. Denitrifying bacteria (more common in anaerobic areas in nature and in denitrification filter units in more advanced culture systems) transform nitrates to nitrogen gas.
6. Plants and phytoplankton (algae) can also use ammonia and nitrates for growth and help remove them from the system water.
7. Most common biofilter bacteria work best between ~ pH 7–8.5.

Ammonia test kits usually measure total ammonia (TA) or total ammonia-nitrogen (TAN). TA (or TAN) includes two forms of ammonia: the more toxic un-ionized ammonia (UIA) and the less toxic ionized ammonia (IA). The fraction of TA that is the more toxic UIA depends on several factors, but the most important are temperature and pH. **Table 2.2** can be used to determine the amount of UIA in a given water body based on the TA, pH, and temperature. To calculate UIA concentration using water test results and **Table 2.2**, use the following steps.

1. Determine if the ammonia test kit provides concentration as "total ammonia nitrogen" (TAN) or "total ammonia" (TA). If the concentration is provided as TAN, multiply the concentration of TAN by 1.2 for TA concentration.
2. To calculate the amount of un-ionized ammonia (UIA) present, multiply the TA by the appropriate fraction number selected from the table using the pH and temperature from your system water sample. If the pH and/or

Table 2.2 Fraction of un-ionized ammonia (UIA) in aqueous solution at different pH values and temperatures. From Francis-Floyd, Ruth, Craig Watson, Denise Petty, and Deborah Pouder. 2022. "Ammonia in Aquatic Systems: FA-16/FA031, 06/2022". EDIS 2022 (4). https://doi.org/10.32473/edis-fa031-2022. Table based on Emerson, K., R. C. Russo, R. E. Lund, and R. V. Thurston. 1975. "Aqueous Ammonia Equilibrium Calculations: Effects of pH and Temperature." *Journal of the Fisheries Research Board of Canada* 32:2379–2383

TEMPERATURE

pH	42.0(°F)	46.4	50.0	53.6	57.2	60.8	64.4	68.0	71.6	75.2	78.8	82.4	86.0	89.6
	6(0C)	8	10	12	14	16	18	20	22	24	26	28	30	32
7.0	.0013	.0016	.0018	.0022	.0025	.0029	.0034	.0039	.0046	.0052	.0060	.0069	.0080	.0093
7.2	.0021	.0025	.0029	.0034	.0040	.0046	.0054	.0062	.0072	.0083	.0096	.0110	.0126	.0150
7.4	.0034	.0040	.0046	.0054	.0063	.0073	.0085	.0098	.0114	.0131	.0150	.0173	.0198	.0236
7.6	.0053	.0063	.0073	.0086	.0100	.0116	.0134	.0155	.0179	.0206	.0236	.0271	.0310	.0369
7.8	.0084	.0099	.0116	.0135	.0157	.0182	.0211	.0244	.0281	.0322	.0370	.0423	.0482	.0572
8.0	.0133	.0156	.0182	.0212	.0247	.0286	.0330	.0381	.0438	.0502	.0574	.0654	.0743	.0877
8.2	.0210	.0245	.0286	.0332	.0385	.0445	.0514	.0590	.0676	.0772	.0880	.0998	.1129	.1322
8.4	.0328	.0383	.0445	.0517	.0597	.0688	.0790	.0904	.1031	.1171	.1326	.1495	.1678	.1948
8.6	.0510	.0593	.0688	.0795	.0914	.1048	.1197	.1361	.1541	.1737	.1950	.2178	.2422	.2768
8.8	.0785	.0909	.1048	.1204	.1376	.1566	.1773	.1998	.2241	.2500	.2774	.3062	.3362	.3776
9.0	.1190	.1368	.1565	.1782	.2018	.2273	.2546	.2836	.3140	.3456	.3783	.4116	.4453	.4902
9.2	.1763	.2008	.2273	.2558	.2861	.3180	.3512	.3855	.4204	.4557	.4909	.5258	.5599	.6038
9.4	.2533	.2847	.3180	.3526	.3884	.4249	.4618	.4985	.5348	.5702	.6045	.6373	.6685	.7072
9.6	.3496	.3868	.4249	.4633	.5016	.5394	.5762	.6117	.6456	.6777	.7078	.7358	.7617	.7929
9.8	.4600	.5000	.5394	.5778	.6147	.6499	.6831	.7140	.7428	.7692	.7933	.8153	.8351	.8585
10.0	.5745	.6131	.6498	.6844	.7166	.7463	.7735	.7983	.8207	.8408	.8588	.8749	.8892	.9058
10.2	.6815	.7152	.7463	.7746	.8003	.8234	.8441	.8625	.8788	.8933	.9060	.9173	.9271	.9389

temperature is between two values, to be more conservative, use the higher pH value and/or higher temperature value to determine the correct fraction number to use.

Example 1

System water pH is 8.2; temperature is 26°C, test kit reads 2.2 mg/L as total ammonia nitrogen (TAN).

1. Convert TAN to TA: 2.2 mg/L x 1.2 = 2.64 mg/L TA.
2. Determine UIA fraction number from **Table 2.2**: pH of 8.2, temperature of 26°C, UIA fraction number is 0.0880.
3. Determine UIA: multiply TA by UIA fraction number to determine UIA: 2.2 mg/L x 0.0880 = 0.194mg/L.

Because UIA should, ideally, be <0.05 mg/L, and the calculated UIA, 0.194 mg/L, is >0.05 mg/L, there most likely will be pathophysiology and gill damage to the fish associated with this harmful concentration.

Example 2

System water pH is 8.1; temperature is 24.5°C; test kit reads 2.2 mg/L TAN.

1. Convert TAN to TA: same as in Example 1 above. 2.2 mg/L x 1.2 = 2.64 mg/L TA.
2. Determine UIA fraction number from **Table 2.2**: the pH and temperature are not on the table, so to be more conservative, use pH 8.2 and a temperature of 26°C. For pH 8.2 and temperature of 26°C, the UIA fraction number is 0.880.
3. Determine UIA: multiply TA by UIA fraction number to determine UIA: 2.2 mg/L x 0.880 = 0.194 mg/L.

Note: although the actual UIA is most likely less than 0.194 mg/L, this conservative method suggests that it still may be high enough to be problematic.

WATER QUALITY: CHEMICAL IMBALANCES AND MANAGEMENT

When testing water, use professional aquaculture test kits (e.g., Hach, LaMotte, and YSI) because these are much more accurate and reliable than most over-the-counter hobbyist kits. Meters (e.g., dissolved oxygen, pH) should be calibrated as per manufacturer's recommendations. Clients' water quality histories are often minimal or unreliable. Verify parameters using a professional test kit, and ascertain whether any water changes or treatments have been provided, since these will affect the interpretation of results.

Water quality testing of the system in question, taken during a disease outbreak, is necessary and will provide good information but does not provide trends.

Consequently, source water should also be tested (after aeration and pre-addition modifications) to provide baseline data that will help provide a system "normal" and can be used to gauge the current trend of the given system as well as track changes over time.

It is also critical to test water as soon as possible during an outbreak before any mitigation has been carried out (water changes, increases in aeration, treatments). Some chemicals will interact with some test reagents and lead to false results. For example, products with formalin and some ammonia-binding products will react with Nessler's reagent (the "yellow" reagent used in some common ammonia tests) and falsely elevate results.

Acidity, Alkalinity, and pH
- Abrupt pH changes are more disruptive than a specific stable pH level.
- Acidity = buffering capacity against strong bases.
- Alkalinity = buffering capacity (not the same as "alkaline") against acid. Sometimes referred to as KH (or "carbonate hardness" because it is a measure of carbonates, bicarbonates, and other acid buffers). Measured in units of mg/L $CaCO_3$. Freshwater alkalinity can vary greatly, but common ranges are 50–250 mg/L. Alkalinity levels can affect the toxicity of some treatments (e.g., copper in freshwater).
- Basic (or alkaline) = pH > 7.0 (do not confuse with "alkalinity").
- Bicarbonate (HCO_3^-), measured as a major component of alkalinity is the primary buffer in water (carbonic acid – bicarbonate buffer system). To increase alkalinity, add buffers: crushed coral, dolomite (1 lb per 10 gallons [453 g per 37.8 L] of water), or commercial products from pet stores.
- Over time, in an aquarium system, organic acids increase, and as ammonia is oxidized to nitrate, alkalinity is used up, and pH decreases. Carbon dioxide can also build up if there is inadequate aeration/degassing. Prevent pH declines with good aeration/degassing and 10% water change every 10 days (q10d) to help replenish alkalinity (assuming source water has adequate alkalinity) but monitor alkalinity and compare to baseline source water to observe trends.
- Emergency acidic/lowered pH water treatment.
  - If the pH has dropped significantly compared to original source water pH due to improper system management:

- Check total ammonia levels first. If ammonia is present, determine what the UIA would be if the pH is elevated to source water (i.e., original) pH, using UIA fraction table and system temperature. If UIA is estimated to be >0.05 mg/L, increasing pH rapidly will shift TA to the more toxic UIA. To prevent this, consider a) adding an ammonia-binding product (available at pet stores or online) prior to attempting water changes, or b) do water changes using water with a pH of 7.0 or below (carefully adjust water source by adding, for example, an appropriate amount of an over-the-counter aquarium pH lowering product).
- After total ammonia has been reduced, aerate water vigorously to expel CO_2.
- If the calculated UIA using the original source water pH and current measured total ammonia is < 0.05 mg/L, gradually change 25%–50% of the water every 48 hours (q48h) until normal and/or increase pH and alkalinity with sodium bicarbonate (baking soda) (**Figure 2.18**).
- Alkaline (more basic/higher pH) water treatment.
  - Filter water through peat; however, peat moss may carry high loads of mycobacteria.
  - Decrease pH with sodium biphosphate or hydrochloric acid (e.g., pool acid).
  - Use commercial products as directed.

Figure 2.18 Sodium bicarbonate (baking soda). RP Yanong

Hardness

- Hardness is also sometimes referred to as GH, or general hardness. It is a measure of total divalent cations (primarily Ca^{2+}, Mg^{2+}, but also Fe^{2+}, Cu^{2+}, Zn^{2+} among others) expressed as mg/L $CaCO_3$.
- Calcium and magnesium are important for growth and development of fish and may be extracted from the water if diet is inadequate. Some species may prefer softer or harder water, although many can acclimate over time, and domesticated/aquacultured species are often more tolerant.
- Hardness, alkalinity, temperature, and pH can affect toxicity and/or effectiveness of some medications.
- Reduce hardness by adding peat or diluting with softened or reverse osmosis water.

Calcium

- For many freshwater fish, a minimum hardness of 50 mg/L is recommended (calcium minimum of 31 mg/L), with Rift Lake African cichlids preferring >180 mg/L.
- Measure in saltwater tanks; 400-450 ppm recommended.
- If using common commercial artificial sea salt brands, calcium concentrations should be sufficient.

Specific Gravity

- Measure with refractometer or hydrometer.
- Freshwater specific gravity may increase to 1.003 when salt is being used as a treatment.
- For conversions from specific gravity to salinity, see **Table 2.3**.

Ammonia

- Any detectable ammonia indicates inadequate biofiltration. This may be due to inadequate time for biofilter bacteria to establish, too many fish and/or overfeeding for the current biofilter set up, some compound (e.g., antibiotics) compromising the biofilter, or a poorly managed system leading to "old tank" syndrome.
- Water quality test kits measure total ammonia, which is a combination of NH_3 and NH_4^+.
- NH_3 is toxic UIA.

Table 2.3 Specific gravity to salinity conversion table for 60°F/60°F hydrometers values taken from tables provided by the LaMotte Company, Chestertown, MD, USA

OBSERVED READING	TEMPERATURE (°C)										
	-1.0	0.0	1.0	2.0	3.0	4.0	5.0	6.0	7.0	8.0	9.0
0.9980	–	–	–	–	–	–	–	–	–	–	–
0.9990	–	–	–	–	–	–	–	–	–	–	–
1.0000	–	–	–	–	–	–	–	–	–	–	–
1.0010	0.6	0.6	0.5	0.5	0.2	0.2	0.2	0.2	0.2	0.2	0.5
1.0020	1.9	1.9	1.6	1.6	1.6	1.6	1.5	1.5	1.6	1.6	1.6
1.0030	3.2	3.1	2.9	2.9	2.8	2.8	2.8	2.8	2.8	2.9	2.9
1.0040	4.4	4.2	4.1	4.1	4.1	4.1	4.1	4.1	4.1	4.2	4.2
1.0050	5.7	5.5	5.4	5.4	5.4	5.3	5.3	5.4	5.4	5.4	5.5
1.0060	6.8	6.8	6.6	6.6	6.6	6.6	6.6	6.6	6.7	6.7	6.8
1.0070	8.1	8.0	7.9	7.9	7.9	7.9	7.9	7.9	7.9	8.0	8.1
1.0080	9.3	9.2	9.2	9.2	9.2	9.2	9.2	9.2	9.2	9.3	9.3
1.0090	10.5	10.5	10.4	10.4	10.4	10.4	10.4	10.5	10.5	10.6	10.6
1.0100	11.8	11.7	11.7	11.7	11.7	11.7	11.7	11.7	11.8	11.8	11.9
1.0110	13.0	13.0	12.8	12.8	12.8	12.8	13.0	13.0	13.0	13.1	13.2
1.0120	14.3	14.1	14.1	14.1	14.1	14.1	14.1	14.3	14.3	14.4	14.5
1.0130	15.4	15.4	15.4	15.4	15.4	15.4	15.4	15.4	15.6	15.7	15.8

(Continued)

Table 2.3 (Continued)

OBSERVED READING	TEMPERATURE (°C)										
	-1.0	0.0	1.0	2.0	3.0	4.0	5.0	6.0	7.0	8.0	9.0
1.0140	16.7	16.6	16.6	16.6	16.6	16.6	16.7	16.7	16.9	17.0	17.0
1.0150	17.9	17.9	17.9	17.9	17.9	17.9	17.9	18.0	18.0	18.2	18.3
1.0160	19.2	19.1	19.1	19.1	19.1	19.2	19.2	19.3	19.3	19.5	19.6
1.0170	20.4	20.4	20.4	20.4	20.4	20.4	20.5	20.5	20.6	20.8	20.9
1.0180	21.7	21.7	21.6	21.6	21.7	21.7	21.7	21.8	22.0	22.1	22.2
1.0190	22.9	22.9	22.9	22.9	22.9	23.0	23.0	23.1	23.3	23.4	23.5
1.0200	24.2	24.2	24.2	24.2	24.2	24.2	24.3	24.3	24.4	24.6	24.7
1.0210	25.3	25.3	25.3	25.3	25.5	25.5	25.6	25.6	25.7	25.9	26.0
1.0220	26.6	26.6	26.6	26.6	26.6	26.8	26.8	26.9	27.0	27.2	27.3
1.0230	27.8	27.8	27.8	27.8	27.9	27.9	28.1	28.2	28.3	28.5	28.6
1.0240	29.1	29.1	29.1	29.1	29.1	29.4	29.4	29.5	29.5	29.8	29.9
1.0250	30.3	30.3	30.4	30.4	30.4	30.6	30.6	30.7	30.8	30.9	31.1
1.0260	31.6	31.6	31.6	31.6	31.7	31.7	31.9	32.0	32.1	32.2	32.4
1.0270	32.8	32.8	32.9	32.9	32.9	33.0	33.2	33.3	33.4	33.5	33.7
1.0280	34.1	34.1	34.1	34.1	34.2	34.5	34.5	34.5	34.7	34.8	35.0
1.0290	35.2	35.2	35.4	35.4	35.5	35.6	35.6	35.8	35.9	36.2	36.3
1.0300	36.5	36.5	36.7	36.7	36.7	36.9	36.9	37.1	37.2	37.3	37.6
1.0310	37.7	37.7	37.8	37.8	38.0	38.1	38.2	38.4	38.5	38.6	38.9

Table 2.3 (Continued)

OBSERVED READING	TEMPERATURE (°C)											
	10.0	11.0	12.0	13.0	14.0	15.0	16.0	17.0	18.0	18.5	19.0
0.9980	–	–	–	–	–	–	–	–	–	–	–
0.9990	–	–	–	–	–	–	–	–	–	–	–
1.0000	–	–	–	–	–	–	0.0	0.2	0.3	0.5	0.6
1.0010	0.5	0.6	0.6	0.7	0.8	1.0	1.2	1.5	1.6	1.8	1.9
1.0020	1.8	1.9	2.0	2.1	2.3	2.4	2.5	2.8	2.9	3.1	3.2
1.0030	3.1	3.2	3.3	3.4	3.6	3.7	3.8	4.1	4.2	4.4	4.5
1.0040	4.4	4.5	4.6	4.8	4.9	5.0	5.1	5.4	5.5	5.7	5.8
1.0050	5.5	5.7	5.8	5.9	6.2	6.3	6.6	6.7	7.0	7.1	7.1
1.0060	6.8	7.0	7.1	7.2	7.5	7.6	7.9	8.0	8.3	8.4	8.5
1.0070	8.1	8.3	8.4	8.5	8.8	8.9	9.2	9.3	9.6	9.7	9.8
1.0080	9.4	9.6	9.7	9.8	10.0	10.2	10.5	10.6	10.9	11.0	11.1
1.0090	10.7	10.9	11.0	11.1	11.3	11.5	11.8	11.9	12.2	12.3	12.4
1.0100	12.0	12.2	12.3	12.4	12.6	12.8	13.1	13.2	13.5	13.6	13.7
1.0110	13.4	13.5	13.6	13.7	13.9	14.1	14.4	14.5	14.8	14.9	15.0
1.0120	14.7	14.8	14.9	15.0	15.2	15.4	15.7	15.8	16.1	16.2	16.3
1.0130	15.8	16.0	16.2	16.3	16.5	16.7	17.0	17.1	17.4	17.5	17.7
1.0140	17.1	17.3	17.5	17.7	17.8	18.0	18.3	18.6	18.7	18.8	19.0

(Continued)

Table 2.3 (Continued)

OBSERVED READING	TEMPERATURE (°C)											
	10.0	11.0	12.0	13.0	14.0	15.0	16.0	17.0	18.0	18.5	19.0
1.0150	18.4	18.6	18.8	19.0	19.1	19.3	19.6	19.9	20.0	20.1	20.4
1.0160	19.7	19.9	20.1	20.3	20.4	20.6	20.9	21.2	21.3	21.4	21.7
1.0170	21.0	21.2	21.3	21.6	21.7	22.0	22.2	22.5	22.7	22.9	23.0
1.0180	22.3	22.5	22.6	22.9	23.0	23.3	23.5	23.8	24.0	24.2	24.3
1.0190	23.6	23.8	23.9	24.2	24.3	24.6	24.8	25.1	25.3	25.5	25.6
1.0200	24.8	25.1	25.2	25.5	25.6	25.9	26.1	26.4	26.6	26.8	26.9
1.0210	26.1	26.4	26.5	26.8	26.9	27.2	27.4	27.7	27.9	28.1	28.2
1.0220	27.4	27.7	27.8	28.1	28.2	28.5	28.7	29.0	29.2	29.4	29.5
1.0230	28.7	28.9	29.1	29.4	29.5	29.8	30.0	30.3	30.6	30.7	30.8
1.0240	30.0	30.2	30.4	30.6	30.8	31.1	31.3	31.6	31.9	32.0	32.1
1.0250	31.3	31.5	31.7	31.9	32.1	32.4	32.6	32.9	33.2	33.3	33.4
1.0260	32.6	32.8	33.0	33.2	33.4	33.7	33.9	34.2	34.5	34.6	34.7
1.0270	33.9	34.1	34.3	34.5	34.7	35.0	35.2	35.5	35.8	35.9	36.2
1.0280	35.1	35.4	35.6	35.8	36.0	36.3	36.5	36.8	37.1	37.2	37.5
1.0290	36.4	36.7	36.8	37.1	37.3	37.6	37.8	38.1	38.4	38.6	38.8
1.0300	37.7	38.0	38.1	38.4	38.6	38.9	39.1	39.4	39.7	39.9	40.1
1.0310	39.0	39.3	39.4	39.7	39.9	40.2	40.5	40.7	41.0	41.2	41.4

Table 2.3 (Continued)

OBSERVED READING	TEMPERATURE (°C)										
	19.5	20.0	20.5	21.0	21.5	22.0	22.5	23.0	23.5	24.0	24.5
0.9980	–	–	–	–	–	–	–	–	–	–	–
0.9990	–	–	–	–	–	–	–	–	–	–	–
1.0000	0.7	0.8	1.0	1.1	0.0	0.1	0.2	0.3	0.5	0.6	0.7
1.0010	2.0	2.1	2.3	2.4	1.2	1.4	1.5	1.6	1.8	1.9	2.0
1.0020	3.3	3.4	3.6	3.7	2.5	2.5	2.7	2.8	2.9	3.1	3.2
1.0030	4.6	4.8	4.9	5.0	3.8	4.0	4.1	4.2	4.4	4.6	4.8
1.0040	5.9	6.1	6.2	6.3	5.1	5.3	5.4	5.5	5.8	5.9	6.1
1.0050	7.2	7.4	7.5	7.6	6.4	6.6	6.7	7.0	7.1	7.2	7.4
1.0060	8.7	8.8	8.9	9.1	7.7	7.9	8.1	8.3	8.4	8.5	8.7
1.0070	10.0	10.1	10.2	10.4	9.2	9.3	9.4	9.6	9.7	9.8	10.1
1.0080	11.3	11.4	11.5	11.7	10.5	10.6	10.7	10.9	11.0	11.3	11.4
1.0090	12.6	12.7	12.8	13.0	11.8	11.9	12.0	12.2	12.4	12.6	12.7
1.0100	13.9	14.0	14.1	14.3	13.1	13.2	13.4	13.6	13.7	13.9	14.0
1.0110	15.2	15.3	15.4	15.6	14.4	14.5	14.8	14.9	15.0	15.2	15.3
1.0120	16.5	16.6	16.7	17.0	15.7	16.0	16.1	16.2	16.3	16.5	16.7
1.0130	17.8	17.9	18.0	18.3	17.1	17.3	17.4	17.5	17.7	17.9	18.0
1.0140	19.1	19.3	19.5	19.6	18.4	18.6	18.7	18.8	19.1	19.2	19.3
1.0150	20.5	20.6	20.8	20.9	19.7	19.9	20.0	20.1	20.4	20.5	20.6
1.0160	21.8	22.0	22.1	22.2	21.0	21.2	21.3	21.6	21.7	21.8	22.0
					22.3	22.5	22.7	22.9	23.0	23.3	23.4

(Continued)

Table 2.3 (Continued)

OBSERVED READING	TEMPERATURE (°C)										
	19.5	20.0	20.5	21.0	21.5	22.0	22.5	23.0	23.5	24.0	24.5
1.0170	23.1	23.3	23.4	23.5	23.6	23.8	24.0	24.2	24.3	24.6	24.7
1.0180	24.4	24.6	24.7	24.8	24.9	25.2	25.3	25.5	25.6	25.9	26.0
1.0190	25.7	25.9	26.0	26.1	26.4	26.5	26.6	26.8	27.0	27.2	27.3
1.0200	27.0	27.2	27.3	27.4	27.7	27.8	27.9	28.2	28.3	28.5	28.6
1.0210	28.3	28.5	28.6	28.9	29.0	29.1	29.2	29.5	29.6	29.8	30.0
1.0220	29.6	29.8	30.0	30.2	30.3	30.4	30.7	30.8	30.9	31.2	31.3
1.0230	30.9	31.2	31.3	31.5	31.6	31.7	32.0	32.1	32.2	32.5	32.6
1.0240	32.2	32.5	32.6	32.8	32.9	33.2	33.3	33.4	33.7	33.8	33.9
1.0250	33.7	33.8	33.9	34.1	34.2	34.5	34.6	34.7	35.0	35.1	35.2
1.0260	35.0	35.1	35.2	35.4	35.6	35.8	35.9	36.0	36.3	36.4	36.7
1.0270	36.3	36.4	36.5	36.7	36.9	37.1	37.2	37.5	37.6	37.8	38.0
1.0280	37.6	37.7	37.8	38.1	38.2	38.4	38.5	38.8	38.9	39.1	39.3
1.0290	38.9	39.0	39.1	39.4	39.5	39.7	39.9	40.1	40.2	40.5	40.6
1.0300	40.2	40.3	40.6	40.7	40.8	41.0	41.2	41.4	41.6	41.8	41.9
1.0310	41.5	41.8	41.9	42.0	42.1	42.3	42.5	—	—	—	—

Table 2.3 (Continued)

OBSERVED READING	TEMPERATURE (°C)											
	25.0	25.5	26.0	26.5	27.0	27.5	28.0	28.5	29.0	29.5	30.0
0.9980	–	–	–	0.1	0.2	0.3	0.6	0.7	0.8	1.1	1.2
0.9990	0.8	1.0	1.2	1.4	1.5	1.8	1.9	2.0	2.3	2.4	2.5
1.0000	2.1	2.4	2.5	2.7	2.9	3.1	3.2	3.4	3.6	3.7	4.0
1.0010	3.4	3.6	3.8	4.0	4.2	4.4	4.5	4.8	4.9	5.1	5.1
1.0020	4.9	5.0	5.1	5.4	5.5	5.7	5.9	6.1	6.3	6.4	6.6
1.0030	6.2	6.3	6.6	6.7	6.8	7.1	7.2	7.4	7.6	7.7	8.0
1.0040	7.5	7.7	7.9	8.0	8.3	8.4	8.5	8.8	8.9	9.2	9.3
1.0050	8.9	9.1	9.2	9.3	9.6	9.7	10.0	10.1	10.2	10.5	10.6
1.0060	10.2	10.4	10.5	10.7	10.9	11.0	11.3	11.4	11.7	11.8	12.0
1.0070	11.5	11.7	11.9	12.0	12.2	12.4	12.6	12.8	13.0	13.1	13.4
1.0080	12.8	13.0	13.2	13.4	13.6	13.7	13.9	14.1	14.3	14.5	14.7
1.0090	14.1	14.4	14.5	14.7	14.9	15.0	15.3	15.4	15.7	15.8	16.1
1.0100	15.6	15.7	15.8	16.1	16.2	16.5	16.6	16.7	17.0	17.1	17.4
1.0110	16.9	17.0	17.3	17.4	17.5	17.8	17.9	18.2	18.3	18.6	18.7
1.0120	18.2	18.3	18.6	18.7	19.0	19.1	19.3	19.5	19.6	19.9	20.1
1.0130	19.5	19.7	19.9	20.0	20.3	20.4	20.6	20.8	21.0	21.2	21.4
1.0140	20.9	21.0	21.2	21.4	21.6	21.8	22.0	22.2	22.3	22.6	22.7

(Continued)

Table 2.3 (Continued)

OBSERVED READING	TEMPERATURE (°C)										
	25.0	25.5	26.0	26.5	27.0	27.5	28.0	28.5	29.0	29.5	30.0
1.0150	22.2	22.3	22.5	22.7	22.9	23.1	23.3	23.5	23.6	23.9	24.0
1.0160	23.5	23.6	23.9	24.0	24.3	24.4	24.7	24.8	25.1	25.2	25.5
1.0170	24.8	25.1	25.2	25.3	25.6	25.7	26.0	26.1	26.4	26.5	26.8
1.0180	26.1	26.4	26.5	26.8	26.9	27.2	27.3	27.6	27.7	27.9	28.1
1.0190	27.6	27.7	27.8	28.1	28.2	28.5	28.6	28.9	29.0	29.2	29.5
1.0200	28.9	29.0	29.2	29.4	29.6	29.8	30.0	30.2	30.4	30.6	30.8
1.0210	30.2	30.3	30.6	30.7	30.9	31.1	31.3	31.5	31.7	32.0	32.1
1.0220	31.5	31.7	31.9	32.0	32.2	32.5	32.6	32.9	33.0	33.3	33.4
1.0230	32.8	33.0	33.2	33.4	33.5	33.8	33.9	34.2	34.5	34.6	34.8
1.0240	34.2	34.3	34.5	34.7	35.0	35.1	35.4	35.5	35.8	35.9	36.2
1.0250	35.5	35.6	35.9	36.0	36.3	36.4	36.7	36.8	37.1	37.2	37.5
1.0260	36.8	36.9	37.2	37.3	37.6	37.7	38.0	38.2	38.4	38.6	38.8
1.0270	38.1	38.4	38.5	38.8	38.9	39.1	39.3	39.5	39.8	39.9	40.2
1.0280	39.4	39.7	39.8	40.1	40.2	40.5	40.7	40.8	41.1	41.2	41.5
1.0290	40.8	41.0	41.2	41.4	41.6	41.8	—	—	—	—	—
1.0300	—	—	—	—	—	—	—	—	—	—	—
1.0310	—	—	—	—	—	—	—	—	—	—	—

Table 2.3 (Continued)

OBSERVED READING	TEMPERATURE (°C)						
	30.5	31.0	31.5	32.0	32.5	33.0
0.9980	1.5	1.6	1.9	2.0	2.3	2.4
0.9990	2.8	2.9	3.2	3.4	3.6	3.8
1.0000	4.1	4.4	4.5	4.8	4.9	5.1
1.0010	5.4	5.5	5.8	5.9	6.2	6.4
1.0020	6.8	7.0	7.2	7.5	7.6	7.9
1.0030	8.1	8.4	8.5	8.8	9.1	9.2
1.0040	9.6	9.7	10.0	10.1	10.4	10.5
1.0050	10.9	11.0	11.3	11.5	11.7	11.9
1.0060	12.2	12.4	12.6	12.8	13.1	13.2
1.0070	13.6	13.7	14.0	14.1	14.4	14.7
1.0080	14.9	15.2	15.3	15.6	15.7	16.0
1.0090	16.2	16.5	16.6	16.9	17.1	17.3
1.0100	17.5	17.8	18.0	18.2	18.4	18.7
1.0110	19.0	19.1	19.3	19.6	19.7	20.0
1.0120	20.3	20.5	20.6	20.9	21.2	21.3
1.0130	21.6	21.8	22.1	22.2	22.5	22.7

(Continued)

Table 2.3 (Continued)

OBSERVED READING	TEMPERATURE (°C)						
	30.5	31.0	31.5	32.0	32.5	33.0
1.0140	23.0	23.1	23.4	23.6	23.8	24.0
1.0150	24.3	24.6	24.7	24.9	25.2	25.3
1.0160	25.6	25.9	26.1	26.3	26.5	26.8
1.0170	27.0	27.2	27.4	27.7	27.8	28.1
1.0180	28.3	28.5	28.7	29.0	29.2	29.4
1.0190	29.6	29.9	30.0	30.3	30.6	30.8
1.0200	30.9	31.2	31.5	31.6	31.9	32.1
1.0210	32.4	32.5	32.8	33.0	33.3	33.4
1.0220	33.7	33.9	34.1	34.3	34.6	34.8
1.0230	35.0	35.2	35.5	35.6	35.9	36.2
1.0240	36.4	36.5	36.8	37.1	37.2	37.5
1.0250	37.7	37.8	38.1	38.4	38.6	38.8
1.0260	39.0	39.3	39.4	39.7	39.9	40.2
1.0270	40.3	40.6	40.8	41.0	41.2	41.5
1.0280	–	–	–	–	–	–
1.0290	–	–	–	–	–	–
1.0300	–	–	–	–	–	–
1.0310	–	–	–	–	–	–

- NH_4^+ is less-toxic IA.
- In general, marine fish are considered more susceptible to ammonia toxicity than freshwater fish, although species-specific sensitivities vary.
- (NH_3: NH_4+) ratio depends on pH, temperature, pressure, and salinity, with pH and temperature most important. See **Table 2.2** to calculate the more toxic UIA (NH_3) concentration based on pH and temperature.
- High pH increases the percentage of more toxic UIA:
  - 3.0 ppm total ammonia at pH 8.5 is deadly.
  - 3.0 ppm total ammonia at pH 6.0 is non-toxic in short-term but stressful.
  - Likewise, higher temperatures will increase levels of more toxic ammonia.
- Reverse high ammonia (>1.0 ppm) with a 30%–50% water change using dechlorinated water every 12–24 hours (q12–24h), but it is critical to monitor UIA levels. An ammonia-binding product may be needed for addition to the system to prevent rise in free UIA if pH increases significantly during the water change.

Nitrite NO_2

- Harmful if >0.1 mg/L, although some species may be more sensitive. Cyprinids, in general, are more tolerant. In marine systems, it is less of a problem because seawater contains a large quantity of chloride, which competes with nitrites for uptake at the gill membrane; however, there are exceptions (e.g., some studies suggest at least one species of marine fish, the red drum, may be more sensitive).
- Nitrites increase methemoglobin, causing respiratory compromise, known as "brown blood disease." Chronic low nitrite levels may lead to anemia.
- Mitigate with 30%–50% water change + 1 g/L (= 0.1%) salt to increase chloride content. Check filtration components and feeding rates and/or move fish to a clean pond or aquarium.
- Susceptibility of pond fish to nitrite toxicity (from most sensitive to least): trout and cool water fish; catfish, tilapia, and striped bass; goldfish and fathead minnows; large/small mouth bass, bluegill, and green sunfish (all members of the Centrarchidae).

Nitrate NO$_3$
- Nitrate is considered much less toxic to most fish, although species and life stage sensitivities vary (e.g., some eggs and fry of rainbow and cutthroat trout die after chronic exposure to levels as low as 1.1–7.6 mg/L). In general, levels >40 mg/L may be stressful; <30 mg/L is recommended; however, for systems with invertebrates (e.g., coral reef systems), much lower nitrate levels are suggested (0–10 mg/L).
- Reduce with regular water changes; in reef systems, denitrification (reducing nitrate to nitrogen gas) may occur deep within live rock; a refugium (a separate but connected tank with lights and plants/algae that consume nitrates) can reduce nitrates, but plants/algae must be harvested periodically.

Chlorine/Chloramine Toxicity
- Chlorinity = amount of Cl, Br, and I (chlorides, bromides, and iodides) dissolved in 1 kg (2.2 lbs) of seawater. Note: chloride is not the same as chlorine, which is highly toxic.
- City water may contain harmful chlorine (0.5–2.0 mg/L) and chloramines (chlorine + ammonia).
- Treatment for chlorine/chloramines toxicity (piping +/- gill necrosis):
  - Bubble 100% O$_2$ into water.
  - Lower water temperature with ice packs to increase dissolved oxygen. For tropical fishes, lower to 70°F (21°C); for goldfish and koi, lower to 55°F (13°C).
  - Add artificial sea salt 1–2 g/L to fresh water.
  - Dexamethasone 2 mg/kg ICe or IV every 24 hours (q24h) for three days.
  - 7 g sodium thiosulfate (water conditioner) removes chlorine up to 2.0 ppm from 1000 L; also neutralizes chloramines but not the ammonia.
  - Some over-the-counter products will neutralize chloramines. Use as directed.

Copper
- More toxic at lower total alkalinity; very toxic to invertebrates (crabs, clams, corals, snails, sea urchins).
- Remove with copper filters (e.g., use of activated carbon).
- Look for copper pipes and city water as source.
- Copper is also used as a treatment for some parasites of freshwater and marine fish but must be dosed and monitored carefully to avoid toxicity (see "Fish Formulary – Anti-Parasitic").

TANK AND WATER TIPS

New Tank Considerations
- Test and balance water chemistry of source water used and keep a record of source water parameters (and re-test source periodically). Many freshwater fish species, especially aquacultured specimens, can acclimate more easily to pH and hardness values outside of their native ranges, but some may be more sensitive.
- If using a biological filter, allow several weeks for bacterial establishment, but to facilitate this, bacteria need a source of ammonia. Add a very small amount of an ammonia source (e.g., ammonium chloride or use a household clear ammonia product [ammonium hydroxide with no additives]) at a concentration of up to 3–4 mg/L ammonia if no fish or invertebrates are used to cycle the filter. One can also add commercial nitrifying bacteria products to help with establishment.
- Add only a few fish at one time.
- Always quarantine new fish for four weeks before adding to main tank. Live plants and invertebrates can also serve as vectors for pathogens, so if possible, quarantine these for two to four weeks prior to addition as well.
- Number of fish: the general guideline of "1 inch of fish per gallon of water (2.54 cm per 3.78 L)," has often been suggested, but in actuality, the number of fish will depend on many factors, including water temperature; species, size, and age of fish; social and behavioral requirements; tank décor, substrate, and aquascaping; filtration and level of water quality management; as well as diet and feeding rate.
- Monitor water quality regularly after addition of new fish, especially TAN, nitrite, temperature, alkalinity, and pH. Dissolved oxygen is typically not an issue early on, but observe for signs of distress and measure if possible.

Establishing a New Biofilter ("Cycling the Tank")
- For new tanks, the biological filter will need to be established, which will require growth of ammonia-oxidizing bacteria and archaea (AOB/AOA) to transform the toxic ammonia (produced by fish and other processes) to nitrites, nitrite-oxidizing bacteria (NOB) to transform toxic nitrites to nitrates, and "Commamox" bacteria (some *Nitrospira* spp.) that may convert ammonia directly to nitrate. This "cycling" can take six to eight weeks, if no supplemental nitrifying bacteria are added to the new system.
- Cycling can be achieved in several ways. For each method described below, until the tank has cycled, ammonia and nitrite levels should be tested routinely (at least two to four times a week).

1. Add ammonia (e.g., ammonium chloride at 3–4 mg/L or a household ammonia cleaner with no perfumes or other additives) after the tank has been set up but before animals are added (check pH). When the ammonia concentration has dropped to zero, continue to add more ammonia, and monitor as ammonia levels drop. Nitrite levels should rise and then fall to zero as nitrate levels begin to rise.
2. The time required to cycle the tank can be reduced by adding a commercial nitrifying bacteria product or by adding some biofilter media and water from a previously established tank that has not had disease issues. Use of media and water from an established tank, while helpful, may, however, potentially introduce pathogens to the system.
3. Add one or two hardy/more tolerant fish to the system and gradually allow the nitrifying bacteria to cycle. Water changes may be required occasionally if ammonia or nitrite levels remain elevated for more than a day or two. Final tank species should be taken into consideration, as should possible biosecurity issues and pathogens occurring with any new fish. Data from some sources (Souza-Bastos et al. 2017) suggest that Oscars (*Astronotus ocellatus*) may tolerate up to 1.3 mg/L UIA for two days, and (*Corydoras schwartzii*) may tolerate up to 0.65 mg/L UIA for two days. Common carp (*Cyprinus carpio*) may tolerate 2.1 mg/L UIA for two days. However, subclinical and cumulative effects of low-level exposure are unknown, so it is critical to minimize the high-end fluctuations and cycle the biofilter as expeditiously as possible. With regard to nitrites, centrarchids (including largemouth bass and bluegill) are the least sensitive to nitrite toxicity, although, after centrarchids, cyprinids appear to be less sensitive to nitrites than other species.

Established Tank Tips
- Tanks that are "topped up" repeatedly (i.e., water is added to counter evaporation only) and not receiving true water changes may accumulate hardness or toxic substances, e.g., copper; these tanks may also build up organic acids, have reduced alkalinity (due to use by the biofilter), and have lowered pH (all part of "old tank syndrome").
- Film on tank surface.
  - White, hard film: lime (calcium carbonate), the result of frequent low water levels.
  - Green spots: green algae growing on pits in the aquarium glass. Remove with a razor blade, or wash tank with diluted bleach prior to use (bleach is toxic to fish; thoroughly rinse and re-establish proper water quality).

Fresh Water Tips
- When moving fish, mix old and new water to allow acclimatization (0.5 pH/hr). A pH difference of 0.5 or less is not usually harmful. If fish have been shipped and/or in a container for more than a few hours, ammonia levels may be elevated, and UIA should be monitored during the acclimation.

Tips for Ponds
- Protect fish from birds, cats, rats, snakes, otters, raccoons, and other wildlife. Some suggested methods include avoiding gradual slopes into the pond, use of strategically placed fishing line above the pond, predator decoys, hide spots (plant cover, caves, or other structures where fish can seek shelter), motion sensors that activate sound or movement to scare off predators, or even use of your pet dog in the backyard.
- Rough estimate for number of fish: maximum four to eight koi, 8–13 inches (20–33 cm) long, per 1,000 gallons (3,785 L). Over time, those fish will increase in size and may breed, greatly increasing biomass in the pond within a few years.
- Have one deep spot to keep water cool in summer; in winter, if the pond freezes, it must have running water and a hole in the ice (fountain or a heater to keep an open spot).
- Ponds with depths as low as 4-6 feet (1–2 m) and poor circulation may become stratified, with lower temperatures and oxygen levels at the deepest levels and warmer temps with more oxygen at the shallower depths. Storms may agitate water and cause mixing of layers with resultant kills due to resulting anoxia, as low oxygen waters deplete overall oxygen levels.
- Goldfish can live through winter, hibernating and living on fat storage (requires ample feeding through autumn).
- High nitrite levels are more frequent in fall and spring with sudden changes in temperature and/or cloudy weather.
  - Check nitrite every 2–3 weeks, and total ammonia nitrogen (NO_2 precursor) weekly (total ammonia nitrogen = NH_3 + NH_4^+). Keep chloride:nitrite ratio (by weight) at least 10:1. Maintain 25–50 ppm Cl in pond, using salt:
  - 10 x pond [NO_2] – pond [Cl] = ppm Cl (if <0, no addition needed).
  - NaCl is approximately 60% Cl.
  - *Note – pond plants vary in their sensitivity to salt. Many can handle 0.05%–0.1% (= 0.5 – 1 g/L), but if you are adding salt for nitrite imbalances and are unsure of plant sensitivity, remove plants prior to treatment.

- Calculating area and volume of ponds and tanks.
  - Calculate surface area with a transit, measuring tape, or pacing.
    - Rectangles or squares: area = length x width.
    - Circles: area = 3.14 x radius^2.
    - Triangles: area = ½ base x height.
  - Volume = length x width x depth; circular tank volume = 3.14 x depth x radius^2.
  - Ponds.
    - Calculate average depth of ponds by taking several depth measurements (if a very large pond, from boat with weighted cord) in a grid pattern.
    - For "natural" rectangular, sloping earthen ponds, actual volume is length x width x maximum depth x 0.4 (correction factor determined by the Army Corps of Engineers, which takes sloping into account).
  - Cubic inches x 0.000579 = cubic feet.
  - Cubic inches/231 = volume in gallons.
  - Cubic feet x 7.48 = volume in gallons.
  - 1 gallon = 3.785 liters.
  - 1 liter = 61.024 cubic inches.
  - 1 cubic foot = 28.317 liters.
  - See USDA Southern Regional Aquaculture Center's "Calculating Treatments for Ponds and Tanks" and "Calculating Area and Volume of Ponds and Tanks" for detailed instructions and conversion tables.

Saltwater Tips

- Ocean water contains a good balance of elements. Artificial mixes simulate seawater but may include higher or lower levels of some elements/trace minerals.
- Salinity: specific gravity of 1.022–1.025 (use hydrometer) or 30–35 g/L (~ ppt or psu) (use a refractometer). Salinity rises with evaporation. Replace evaporative loss with dechlorinated fresh water, or, ideally, for evaporation in marine systems/reef tanks, use reverse osmosis water to top off. When replacing salt water during water changes, use commercial preparations (mix before adding) or seawater not table salt.
- Light: marine fishes and invertebrates (e.g., anemones, coral) require more light than freshwater species. Leave lights on for 6–12 hr/day.

- Substrate: crushed coral or dolomite ($CaMg(CO_3)_2$) > 2mm.
- Reef systems have more stringent water quality, lighting (intensity and spectra), and water flow requirements than fish-only aquaria.

SALT CALCULATIONS FOR COMMON VOLUMES OF WATER

Types of salt (NaCl-sodium chloride) that can be used include coarse-grain, meat-curing grade, water-softener grade, "solar salt" with no additives, and artificial sea salt (see **Figures 2.19–2.20**). NaCl is approved (Food and Drug Administration (FDA)-Generally Regarded as Safe (GRAS) declaration) for use on food fish as an osmoregulatory enhancer. Salt by its osmoregulatory action causes fish to release large amounts of mucus from their skin and gills. The release of mucus removes and/or kills (at high salt concentrations) some external parasites on the fish. At low concentrations, salt reduces osmotic stress during handling, holding, and hauling.

Suggested concentrations and indications for use of NaCl (also see "Fish Formulary – Anti-Parasitic"):

1. For many freshwater fish species, 10–30 g/L (ppt) as a protist and monogenean parasite treatment (for 5–30 minutes or until fish show signs of stress). Some species, including those with electrosensory capabilities (e.g., mormyrids like the elephantnose fish) may be highly sensitive and unable to tolerate even 1 g/L or less.
   a. For 10–30 g/L NaCl, add the following, based on water volume:
      i. 10 gallons (~38L): 380–1140 g (= 0.84–2.51 lbs).
      ii. 100 gallons (~380L): 3800–11,400 g (= 8.4–25.1 lbs).

Figure 2.19   Solar salt. RP Yanong

Figure 2.20 A 1500 pound (681 kg) bag of commercial sea salt. GA Lewbart

2. For prolonged (12–48 hours or more) stress relief or transport, 1–2 g/L NaCl, add the following based on water volume:
   a. 10 gallons (~38 L): 38–76 g.
   b. 100 gallons (~380 L): 380–760 g.
3. For transport, 0.1–0.2 g/L NaCl may be helpful.
   a. 10 gallons (~38L): 3.8–7.6 g.
   b. 100 gallons (~380 L): 38–76 g.

CHAPTER 3
NUTRITION AND DIET

GENERAL NUTRIENT REQUIREMENTS AND FEED INGREDIENTS

- In the wild, fish species vary widely in their diet. In human care, many common aquarium fish can be considered functional omnivores, although required levels of good quality fat and protein will vary by species, life stage, and reproductive status. In general, fish in human care are relatively less active and can have greater access to food than their wild counterparts, so both quality and quantity should be evaluated. A number of good quality commercial diets may be adequate for more commonly kept species.
- More grossly visible clinical signs, e.g., wasting or emaciation (see **Figures 3.1 and 3.2**) or coelomic distension (if nutrition-related) may be the result of an inappropriate diet that is not bioavailable or balanced, inappropriately apportioned (too little or too much or fed at the wrong time), or poorly stored leading to breakdown or contamination. A zebrafish body condition scoring (BCS) can be approximated or extrapolated for use in other species (see **Nutrition Figure 3.3 and 3.4**) and can be used for periodic assessment to help inform client feeding practices.

MACRONUTRIENTS

- Carbohydrates are not required by many fish (although goldfish and koi use hindgut fermentation to digest complex carbohydrates, and some species have other specializations; other fishes excrete plant matter undigested).
- Most energy is from dietary fat (not more than 15% of the daily intake to avoid hepatic lipidosis).
- Most commercial diets list protein as the largest percentage of dry matter in the feed. Excess protein breaks down into toxic ammonia and nitrites. Animal proteins are most desirable, as they contain the ten essential amino acids. Plant-derived proteins may require balanced supplementation with lysine, methionine, and threonine.

DOI: 10.1201/9781003057727-3

Figure 3.1 The lateral view of a very thin goldenback triggerfish (*Xanthicthys caeruleolineatus*). Note the concavities in the epaxial and abdominal areas. GA Lewbart

MICRONUTRIENTS/MINERALS

- **Table 3.1** and **Table 3.2** provide a summary of potential clinical signs caused by vitamin and mineral deficiency/imbalance.
- Goldfish and koi are cyprinids and, therefore, lack a true acid stomach; consequently, calcium and phosphorus in fish meal are not bioavailable to them; they require inorganic forms of phosphorus (monobasic) and must be fed goldfish- and koi- (i.e., cyprinid) specific diets. Pipefishes and parrotfishes are also agastric.

ADDITIONAL CONSIDERATIONS

- Form: flakes, floating pellets, sinking pellets, sinking wafers, freeze-dried, frozen, live. Store flakes and pellets in cool, dry, dark areas (follow manufacturer recommendations) to reduce degradation; check expiration dates; replace every six to eight months. Storing flakes or pellets in the refrigerator or freezer may help reduce degradation, but it is recommended to subdivide into smaller quantities that can be removed periodically to minimize condensation that may result from continued removal and replacement of the entire batch.
- Nutrient leaching: water-soluble vitamins are lost in water from some flake diets within 30 seconds. Pelleted or granular feeds lose less by leaching.
- Provide variety to avoid malnutrition, even with commercial diets.
- Add fresh vegetables when possible. Diet in the wild can help inform.

Figure 3.2 The dorsal view of a very thin goldenback triggerfish (*Xanthicthys caeruleolineatus*). Note the concavities in the epaxial areas. GA Lewbart

- Avoid live fish as food, as this increases the risk of disease/parasite transmission. If unavoidable, consider quarantining and evaluating live fish prior to feeding.
- Brine shrimp (*Artemia*) are fairly nutritious at early life stages (one to two days post hatch) but do not provide enough protein or other nutrients when

Adult Zebrafish BCS		
	Lateral View	Dorsal View
BCS 1: • Head larger than body (big head) • Lateral- concave ventral surface between head and abdomen (narrow abdomen) • Dorsal- body is more narrow than head and linear • **Fish is thin (emaciated)**		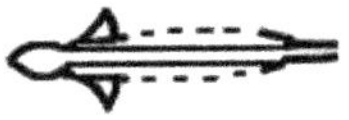
BCS 2: • Head and body equal size • Lateral- flat ventral surface between head and abdomen • Dorsal- head and width of abdomen are equal • **Fish is underconditioned**		
BCS 3: • Body larger than head • Lateral- slight convex ventral surface • Dorsal- head is slight smaller to a fusiform body • **Fish is well-conditioned**		
BCS 4: • Body significantly larger than head • Lateral- body moderately convex ventral surface • Lateral- Symmetrical ventral surface • Dorsal- head visually smaller to a moderately distended abdomen • **Fish is over-conditioned**		
BCS 5: • Body significantly larger than head • Lateral- body significantly convex ventral surface • Lateral- Symmetrical or asymmetrical ventral surface • Dorsal- head visually smaller to a significantly distended abdomen • **Fish is obese (large)**		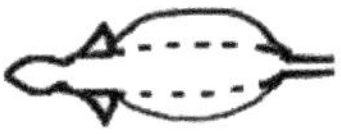

Figure 3.3 Adult zebrafish (*Danio rerio*) body condition scoring chart. Courtesy of the American Association for Laboratory Animal Science

frozen and then thawed. Don't use it as the only feed. Brine shrimp supplements (i.e., Selco or other products added to live brine shrimp in water that they will uptake) may be helpful with older, live brine shrimp to enhance their nutritional value.

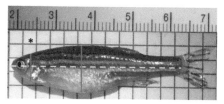 

Figure 3.4 Adult zebrafish (*Danio rerio*) diagrams showing BCS measurement locations. Courtesy of the American Association for Laboratory Animal Science

AMOUNT

- A common general guideline is to feed once daily, no more than fish can eat in 3–5 minutes; however, ideally, determining/estimating the biomass of fish and using a specific percent daily biomass feeding (1%–3% weight of fish for adult maintenance) is preferred. Different diets and feed types (e.g., flake vs. pellet) may "appear" similar in amount when fed but differ radically in actual macronutrient density. Regular observation of a fish's body condition and overall appearance and behavior will help determine if the diet and feeding rate is adequate.
- Knowledge of a species' normal appearance will help when assessing body condition, but often very thin fish will appear to have a much larger head relative to body compared to healthy fish of the same species and reduced mass in the dorsolateral musculature and/or bony structures in the head will be more prominent and fish will appear to have a "sunken in" coelom/abdomen (**Figures 3.1–3.2**). A dorsoventral view of the fish may highlight this.
- Juvenile fish may require multiple daily feedings and typically have a higher daily percent body weight feeding requirement (but use clinical observations and judgment).
- Mature warm water ornamentals (e.g., neon tetras, zebra danios, cichlids, gouramis) require 1.0%–2.5% of their body weight in food per day.
- Goldfish maintained at 68°F (20°C) require 0.3% of their body weight per day in food.
- "Neighbor syndrome": neighbors overfeed fish while owners are away, causing ammonia levels to rise. Avoid by not feeding when gone; most fish can readily survive 7–10 days without supplemental food as long as prior nutrition has been adequate and water quality is good. Of course, if there is the potential in multi-species tanks for predation of smaller fish by larger fish, feeding during this time period may be necessary.
- Another option is to prepackage food into daily meals for proper allocation.

Table 3.1 Important Vitamins for Fishes and the Clinical Signs Related to a Deficiency

VITAMIN	DEFICIENCY SIGNS
A	Exophthalmia, eye-lens displacement, edema, ascites, corneal thinning and expansion, degeneration of retina, gill and skin hemorrhages, skin depigmentation, opercular malformation.
D	Tetany of white skeletal muscle, impaired calcium homeostasis, low bone ash.
E	Skin depigmentation, hemolytic anemia, ascites, immature variable-sized erythrocytes, muscular dystrophy, exudative diathesis, splenic and pancreatic hemosiderosis.
K	Prolonged blood clotting, anemia, lipid peroxidation, reduced hematocrit, skin hemorrhages.
Thiamine (B1)	Muscle atrophy convulsions, instability and loss of equilibrium, edema, darkening. Thiaminases, found in the tissues of certain freshwater and marine species, can hydrolyze thiamine if in contact for prolonged periods before consumption.
Riboflavin	Corneal vascularization, cloudy lens, hemorrhagic eyes, skin and liver hemorrhages, photophobia, dim vision, lack of coordination, abnormal pigmentation of iris, striated constrictions of abdominal wall, dark coloration, anemia, short body dwarfism.
Pyridoxine (B6)	Neurologic disorders, ascites, anemia, rapid postmortem rigor mortis, rapid and gasping breathing.
Pantothenic acid	Clubbed gills, prostration, necrosis and scarring, cellular atrophy, gill exudate, sluggishness, anemia, skin erosion/hemorrhages.
Inositol	Distended stomach, increased gastric emptying time, skin lesions.
Biotin	Lesions in colon, skin depigmentation/lesions, muscle atrophy, spastic convulsions, hemolysis/anemia.
Folic acid	Lethargy, fragility of caudal fin, dark coloration, macrocytic anemia.
Choline	Poor food conversion, fatty liver, hemorrhagic kidney and intestine.
Nicotinic acid/niacin	Anemia, skin and fin lesions +l - hemorrhages, exophthalmia, lesions in colon, jerky or difficult motion, weakness, edema of stomach and colon, muscle spasms while resting.
Cyanocobalamin (B12) ascorbic acid (C)	Low hemoglobin, hemolysis, macrocytic anemia, scoliosis, lordosis, impaired collagen formation, deformed cartilage, eye lesions, internal and external hemorrhages.
p-Aminobenzoic acid	No abnormal indication in growth, appetite, mortality.

Table 3.2 **Important Minerals for Fishes and the Clinical Signs Related to a Deficiency**

MINERAL	IMPORTANCE	REQUIREMENT	DEFICIENCY SIGNS
Calcium (Ca)/ Phosphorus (P)	Calcium comprises 0.5%–1.0% of fish, whole wet weight, with Ca:P ration 0.7–1.6. As much as 99% of Ca and 85%–90% P in skeleton and scales (20–40% in scales). Other functions, e.g., blood clotting and muscle and nerve function (Ca), and membrane structural components, ATP, DNA, RNA (P), pH buffering.	Ca requirement: Most met through uptake from water, 0.45% in calcium-free water (channel catfish). Phosphorus requirements: (as percent of diet) 0.45% (channel catfish), 0.6% (common carp), 0.9% (Nile tilapia). Note: Some forms of phosphorus, e.g., phytate phosphorus, comprising 67% of P in grains, not well utilized by fish. Fish meal P is more available to fish with stomachs than to cyprinids, such as koi, which lack stomachs (25% available). Inorganic P (sodium or monocalcium phosphate) most efficient source.	Deficiencies: Reduced growth (Ca and P); poor bone mineralization, increases in carcass fat, reduced blood phosphate levels, decreased glycogen, and deformities of head (frontal bone), ribs, soft rays of fins, and spine.
Magnesium	Important structural and enzymatic component; muscle function. Diets with plant material usually more than adequate.	0.025%–0.07% of diet (freshwater fish); 0.04% (channel catfish); 0.04%–0.05% (common carp); 0.59–0.77 g/kg (tilapia).	Poor growth, anorexia, lethargy, flaccid muscles, increased mortality, convulsions, spinal deformities, degeneration of epithelial cells of pyloric cecae and gills.
Iron	Important for hemoglobin structure and function. Can be absorbed via water, unless precipitates out of solution. Oral supplementation if feed is primarily of plant origin. Bioavailability higher in animal feed components.	30 mg/kg diet (channel catfish); 150 mg/kg (common carp).	Microcytic anemia.

(Continued)

Table 3.2 *(Continued)*

MINERAL	IMPORTANCE	REQUIREMENT	DEFICIENCY SIGNS
Copper	Iron uptake/utilization; hematopoiesis; collagen synthesis.	3 mg/kg diet (common carp);1.5–5 mg/kg (channel catfish); 3–4 mg/kg (tilapia); 32 mg/kg growth depression/anemia (channel catfish);100–250 mg/kg toxic.	Reduction in iron levels.
Iodine	With tyrosine, thyroid hormone synthesis; absorbed in water and feed (fish meal).	1–5 mg/kg diet.	Goiter (thyroid hyperplasia).
Zinc	Enzyme function (carbonic anhydrase, protease), prevention of keratinization of carbohydrate metabolism.	20 mg/kg (channel catfish); 15–30 mg/kg (common carp); 10 mg/kg (tilapia).	Depressed growth, increased mortality, anorexia, skin/fin erosions, lens cataracts.
Manganese	Amino acid and fatty acid metabolism, glucose oxidation.	2.4 mg/kg (channel catfish); 12–13 mg/kg (common carp and rainbow trout); 12 mg/kg (tilapia).	Depressed growth, abnormal tail growth, stunting, increased mortality.
Selenium	Component of glutathione oxidase; acts with vitamin E as an antioxidant and cofactor for glucose metabolism.	0.38 mg/kg (rainbow trout); 0.25 mg/kg (channel catfish); toxicity 15 mg/kg (channel catfish).	Muscular dystrophy (+ / - vitamin E deficiency).
Sodium, potassium, chloride	Numerous functions, including osmoregulation, pH maintenance.		Rare.

CHAPTER 4
GENERAL REPRODUCTION

SEXING SELECT SPECIES/GROUPS

Reproductively mature individuals often differ from juveniles in appearance and behavior and by sex. Knowledge of sexual dimorphism and dichromatism (sexual differences in shape and coloration) provides important social and biological contexts, including increased territoriality and aggression or normal sexual differences in external appearance. If males and/or conditions are not amenable for spawning, mature females may become "egg bound."

Some freshwater ornamental fish species can be sexed more readily than others. Some of these are described below. Marine species vary in their sexual differentiation, with some having juvenile and intermediate morphology and coloration, and many are difficult to sexually differentiate until they are mature and reproductively active.

Cichlids (Cichlidae)
- Age 2–24 months before sexually mature.
- Females tend to be larger; have a slightly rounder abdomen, shorter dorsal and anal fins; lack ray extensions on any of the fins; and are less colorful. Males may have elongated fins or extended rays on the fins.
  - The females of several species of new world cichlids have a spot in the middle of their dorsal fin.
  - Males of many African Rift Lake cichlids have pseudo-ocelli/"egg spots" on their anal fin.
  - A number of new and old world cichlids are sexually dimorphic and/or dichromatic (e.g., *Parachromis* spp.).
- Genital papilla differences may also be used, although these may be more obvious in some species than others and are much clearer during breeding season (**Figures 4.1–4.3**). Females have a larger opening on their papillae (ovipositor (OV)) for egg deposition.
- Many differences among species.
- Medium to large Central and South American cichlids (including angelfish and discus) and dwarf cichlids spawn on surfaces (slate, plants, PVC) and protect the eggs.

DOI: 10.1201/9781003057727-4

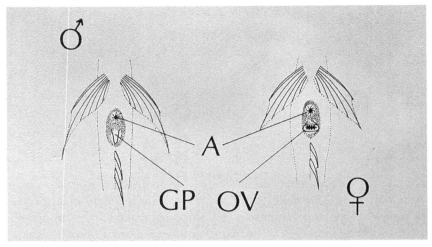

Figure 4.1 Diagram of how to sex a cichlid based on urogenital papillae. Male is on the left, female on the right. A = anus; GP = genital papilla; OV = ovipositor. Courtesy of P Loiselle

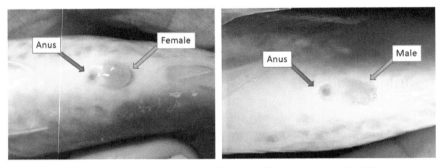

Figure 4.2 Photos of how to sex a red dovii cichlid (*Parachromis dovii*). RP Yanong

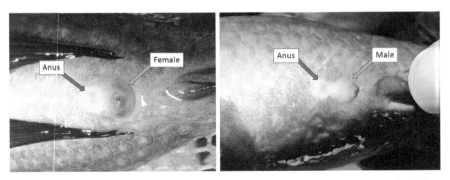

Figure 4.3 Photos of how to sex a three-spot cichlid (*Cichlasoma trimaculatum*) – RP Yanong

- Common Rift Lake African cichlids are mouthbrooders, guarding eggs and larvae within their mouths.

Common Livebearers (Poeciliidae)
- Easier for beginners to breed (guppies, mollies, platies).
- Livebearing fish must be 3–4 months old before breeding. Mature males develop a modified anal fin (gonopodium) that is tube shaped for transfer of sperm packets (spermatozeugmata) to the female (see **Figure 4.4**). After the gonopodium has developed, linear growth ceases (later maturing males tend to be larger than early males). Immature males and females have a more fan-shaped anal fin (see **Figure 4.4**). Mature females that have bred with a male also typically have a dark "gravid spot" in the caudal portion of their body cavity. In some species, females can store sperm for weeks to months.

Gouramis and Bettas (Osphronemidae)
- Male gouramis and male bettas are bubble nest builders and, after spawning, will guard these nests aggressively, even against the female with which they just bred.
- Male gouramis are often smaller and slimmer than females, but more distinctly, male gouramis have a longer and more pointed dorsal fin (see **Figure 1.28**) than females, which have a shorter and more rounded dorsal. In dwarf gourami species, the male is also much more colorful than the female.
- Male bettas are well-known to be aggressive toward other conspecific males. Male bettas have much longer and elaborate finnage (see **Figure 4.5**)

Figure 4.4 Swordtail (*Xiphophorus hellerii*) sexing. Note the gonopodium – the modified anal fin – used for copulation in the male (yellow arrow) and the more triangular anal fin of the female (blue arrow). Courtesy FTFFA

Figure 4.5 Male halfmoon betta (*Betta splendens*) with characteristic large, elaborate finnage seen in this sex. Courtesy of 5-D Tropical Inc.

Figure 4.6 Male (top center – large dorsal fin and more vibrant coloration) and female (lower right – more muted coloration and less elaborate finnage) featherfin rainbowfish. Courtesy Valley Fisheries Inc.

compared to females, which have much shorter fins and often more drab coloration.

Rainbowfishes (Melanotaeniidae)
- Male rainbowfishes, especially during breeding season, are much more colorful than their female counterparts (**Figure 4.6**). Male rainbowfish also tend to be deeper bodied ("taller") and larger than females. Immature rainbowfish are less colorful.

Goldfish and Koi
- Sexually mature and conditioned goldfish males develop nuptial tubercles (koi do not) on their opercula and pectoral fins (**Figure 4.7**). These

General Reproductive Strategies

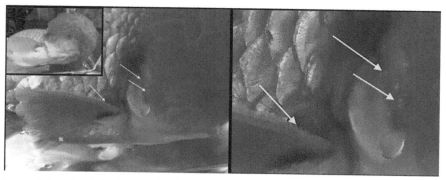

Figure 4.7 Nuptial tubercles (yellow arrows) on a male oranda goldfish. GA Lewbart

tubercles resemble small bumps or nodules that may be mistaken for lesions. Otherwise, mature and conditioned females (holding eggs) will be more robust with enlarged coeloms.
- Koi and goldfish typically spawn during the spring when temperatures near 68°F (20°C). Females spawn "sticky" eggs in vegetation (or vegetation like artificial substrates) that are immediately fertilized by the males. Reproductively conditioned and mature fish may also be induced to spawn with hormones including HCG (trade name Chorulon®) and GnRHa and domperidone (trade name Ovaprim®), with ovulation occurring approximately 12 hours after injection.

Physiological Quick Facts

Life span: freshwater tropical fish live for several years, some up to 20 years.

Body temperature: equal to water temperature.

Heart rate: 30–100 beats/min, variable with species.

Respiratory rate: 40–80 breaths/min, variable with species.

GENERAL REPRODUCTIVE STRATEGIES

Freshwater Egg Layers with No Parental Care
- Type 1 egg scatterers: tetras (Characidae), cyprinids (barbs, goldfish, koi, danios, rasboras), *Corydoras* catfish, and many others. Often differences in sexes, especially out of spawning season, are subtle, but in season, males are

often more slender/elongate, whereas gravid females have a more "rounded" or "fuller" coelom. Nutrition and water quality (e.g., pH, hardness, alkalinity, temperature, total dissolved solids (TDS)), appropriate spawning substrate (non-toxic yarn mops, plants, moss, or hard substrate) are important for reproductive success. In 12 to 48 hours, the eggs hatch, and yolk sac fry emerge which then feed off their yolk sacs for about 4 to 7 days until they are free swimming, at which point they can be properly fed. Goldfish can grow to 1 inch (2.5 cm) in 1 month.
- Type 2 egg scatterers: including (Cyprinodontidae) killifish, and (Melanotaenidae) rainbowfish. Males are often more colorful with more elaborate finnage. Many species of killifish are very short-lived, living only about a year. In the wild, these "annual" killifish evolved to live, reproduce, and die in temporary bodies of water during wet seasons. Eggs of these fish go through a resting phase, or "diapause," completing development when the wet season returns. Hobbyists often use peat as a substrate for these fish, and, after spawning, the peat is gently squeezed to keep it damp but not wet. This egg-laden peat is then kept in a tightly closed container for 6 weeks to 9 months (based on species) prior to release into a shallow container of water for hatching. Non-annual killifish do not require this elaborate protocol. These killifish are often spawned in trios of one male and two females, using floating or weighted mops. Rainbowfishes, predominantly from Australia and New Guinea, condition best with a diet containing some live foods (e.g., brine shrimp). They can be spawned as a pair or group of two to three females per male and, in human care, scatter their eggs in spawning mops or Java moss. Females lay a small number of eggs daily for several weeks, and eggs hatch in approximately 1–3 weeks.

Freshwater Egg Layers with Parental Care
- Labyrinth fish have an increased degree of parental care. The Osphronemidae, which includes most of the common gouramis, bettas, and paradise fish, either build bubble nests (most) or are mouthbrooders (e.g., chocolate gourami, some betta species). Males build bubble nests in which to place and care for the fertilized eggs and defend these nests fiercely – even against the female. Male bettas, after spawning, are sensitive to any environmental perturbations and may eat their eggs if disturbed.
- Cichlids are incredibly diverse with regard to adult size, color, morphology, behavior, and reproduction. Two major reproductive strategies into which cichlids can be divided are monogamy (pair spawning) and polygamy (group

spawning, multiple partners). Although there are always exceptions, common Central and South American cichlids (including angelfish, discus, Jack Dempseys, and oscars) are pair spawners that lay eggs on a substrate and guard them. Most common Rift Lake African cichlids are polygamous and mouthbrooders, with one or both sexes holding fertilized eggs and larvae in their mouths for protection. Dwarf African and South American cichlids, such as the *Apistogramma* spp., spawn with one male and multiple females. Cichlids are extremely territorial and even more so when reproductively active.

Livebearers (Poeciliidae)

- As discussed, males of the common livebearers (poeciliids) have a modified anal fin, the gonopodium, which is used to insert sperm packets (spermatozeugmata) into the female reproductive tract. Depending upon the species and individual, this may be used to immediately fertilize mature oocytes or may be nourished and sequestered in the ovarian cavity wall and used to fertilize multiple batches of eggs. Guppies, platies, swordtails, and variable platies have a gestation period of 4–6 weeks with an average of 50–100 fry. Mollies may require 6–10 weeks.

COMMON MARINE SPECIES

Marine fishes, in general, have more complex reproductive strategies than common freshwater fish species. Although many species are gonochoristic – meaning individuals will become, as adults, either male or female – a number of marine species are functional hermaphrodites, having both mature male and female gonads, either at the same time or becoming one sex and then another during different reproductive cycles. Synchronous (simultaneous) hermaphrodites may have mature testes and ovaries and, within the same reproductive cycle, alternate roles (e.g., *Hypoplectrus* spp. (hamlets) and at least one species of moray, the geometric moray eel (*Gymnothorax griseus*)). Protandrous hermaphrodites develop as mature males first and then may change to female (e.g., clownfishes) based on social and environmental cues. Protogynous hermaphrodites are females first and later may develop into males (*Anthias* spp., wrasses, angelfishes, and parrotfishes). Some species are known to have the ability to switch back and forth between sexes (dottybacks, cleaner wrasse, some marine angelfishes). These changes may explain differences in social behaviors and structure in an aquarium setting over time.

GENERAL NOTES ON TROPICAL FISH EARLY LIFE STAGES

Freshwater Livebearers

- When livebearers are born, they sink to the tank bottom briefly before swimming and can eat when born. They should be separated to avoid cannibalism by larger fish; move fry to tank with other fish once they grow too big to be consumed by other fish.

Freshwater Tropical Egg Layers

- Move fertilized eggs (from spawning substrate), if possible, into a separate aerated container for hatch (to avoid cannibalism by larger fish).
- For outside spawned fish, fry survival in ponds is poor unless intensively managed; water beetles and beetle or other insect larvae may eat fish fry in ponds.
- Use filtration that will not harm eggs or hatching fry.
- For many common tropical freshwater species, eggs hatch over 1–2 days; earlier hatching fry are larger and may eat others.
- Immediately post hatch, larvae often survive on their yolk sac for 1–2 days prior to requiring exogenous food.
- Once they are feeding, food provided must be small enough for fry to consume. Infusoria (microscopic organisms, including free-living, non-parasitic ciliates), rotifers, or microworms may be cultured by more advanced hobbyists to feed these stages. Some fry may eat commercially available powdered "fry foods" and newly hatched brine shrimp (*Artemia* sp.; <24 hours old).
- Split the spawn into smaller, more manageable groups as they grow.
- For more reproduction related troubleshooting information, see Yanong, 1996.

Marine Fishes

Many commonly kept tropical marine fishes have not yet been commercially bred. For those that are aquacultured, many have a very small, prolonged (~2–7 weeks) pelagic phase in which the fish larvae are planktonic ("free-floating") and have not yet settled or fully developed into their adult form. During this pelagic phase, larvae must be fed small live feeds, including microalgae and copepods or rotifers. Breeding and raising marine fishes – except for clownfishes (which have shorter pelagic phases) – are usually beyond the scope of most hobbyists.

EGG BINDING (DYSTOCIA)

Egg binding (dystocia) is one common differential for prolonged coelomic distension in reproductively mature female fish (See **Figure 5.1D**). True egg binding cases may be caused by lack of proper environmental (e.g., photoperiod, temperature, substrate, pH) or biological cues (e.g., lack of males). Other non-reproductive causes of coelomic distension (neoplasia, gas, etc.) should be ruled out, and egg binding may also be the result of an infectious disease.

Knowledge of reproduction and ovulation triggers for that species, presence of males, and use of ultrasound or ovarian biopsy can all help support the diagnosis of egg binding. An ovarian/oocyte biopsy can be performed by inserting a small gauge piece of Silastic™ or similar silicon tubing (for koi, ~ 1.57 mm inner diameter [ID] and 2.41 mm outer diameter [OD]) into the urogenital opening (e.g., on urogenital papilla of females), threading it toward the ovary, and using gentle suction (mouth or syringe) to obtain a small sample of oocytes/eggs to determine condition and maturation status (see **Figure 4.8A**). Smaller or larger ID and OD may be required for other species. If the biopsy reveals degraded or atretic oocytes, or smaller oocytes with central nuclei, these fish are not ready to spawn/ovulate (see **Figure 4.8B**). Mature eggs (oocytes) will be numerous, intact, large, uniform, and with a more peripherally located ("eccentric/off-center") nucleus (aka germinal vesicle), which indicates that the fish is close to ovulation (see **Figure 4.8C**). For marine fish, the nucleus/germinal vesicle may be difficult to see without use of a clearing solution. For naturally opaque eggs, Serra's fixative (ethanol, formalin, glacial acetic acid at a ratio of 6:3:1 by volume for 1–2 minutes) is a good clearing solution protocol and will allow visualization of the germinal vesicle (nucleus). In the absence of a clearing solution, visualization of primary growth oocytes (clear, small) vs. much larger, darker oocytes of

Figure 4.8A Catheterization of koi female for ovarian/oocyte biopsy. A section of small bore, flexible silicone rubber tubing/catheter is placed into the gonoduct to access the ovary. With gentle suction using a syringe (or mouth), a small sample of oocytes can be collected for microscopic staging of oocyte maturation. UF IFAS Tropical Aquaculture Laboratory.

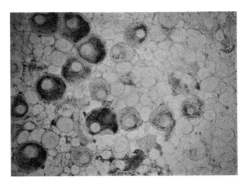

Figure 4.8B Mixed oocyte stages in this sample from a gourami: predominantly immature primary growth oocytes ("clearer" and smaller oocytes) and a few vitellogenic oocytes (larger, with dark gray/brown "granular" cytoplasm). Not a good candidate for hormone induced spawning. M Hyatt and J Meegan.

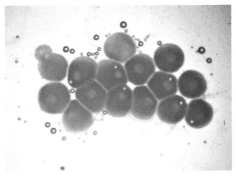

Figure 4.8C Relatively uniform, larger, mature oocytes of a rainbow shark (cyprinid) female with eccentric nuclei ("starred" eggs indicate migration of nucleus toward the periphery), ready for ovulation (and for hormone induced spawning). A Wood

uniform size may provide some evidence of maturation stage. If eggs are mature with peripheral nuclei, natural spawning can be attempted with known environmental or other cues. Otherwise, induced ovulation may be attempted using spawning aids Chorulon (HCG) or Ovaprim (GnRHa + domperidone) as per label instructions (see FORMULARY in Chapter 14). After the drug manufacturer's suggested waiting period after injection (often 4–12 hours), some fishes may release oocytes on their own. However, if necessary, gentle massage of the coelomic cavity from anterodorsal to posteroventral (along the ovaries) toward the gonoduct/genital papilla may result in expression of ovulated eggs.

CHAPTER 5

CLINICAL CASE WORK UPS

INITIAL CONSIDERATIONS

FIRST VISIT/ANNUAL EXAMINATION PROTOCOL

Site Visit

Initial site visits are preferred when feasible. History and in-person initial evaluation of the system and management are much more accessible, and onsite examinations often reveal potential issues that may have been overlooked by an office visit alone. However, there may be limitations on work up, depending on the problem and access and mobility of relevant diagnostic equipment and supplies.

Clinic Visit with Virtual Site Visit

A good alternative is a virtual site visit, ensuring clear real-time observations of fish, systems, equipment, water testing supplies, and feed, followed by a clinic visit.

Fish brought to clinic: use a covered, insulated container, e.g., a cooler, only one-third filled with water. Battery-operated baitfish aerators can be used to oxygenate the water in transit. Have client bring in a separate water sample in a clean, well-rinsed container (e.g., a plastic drinking water bottle rinsed several times first with system water) on ice, any medication being used, and food being fed or good quality photos of these including labels and expiration dates (if not seen during the virtual visit).

History Questions
- In general, history questions should cover:
  - The problem, the affected fish's clinical signs, and other fish and animals in the same system.
  - System filtration and tank/pond specifics, water source and water quality, and general system management protocols.
  - Attempted/past treatments, historical disease problems, and general disease management approaches.
- Infectious agents can enter the system from:

DOI: 10.1201/9781003057727-5

- Feed.
- Water.
- Newly introduced fish.
- Fomites (contaminated nets or other equipment).
- Vectors (including snails, someone's hands or arms if they have been working with another aquarium or fish, plants, or, for outdoor ponds, birds, small mammals, aquatic reptiles).
* Some relevant questions include:
    - What is the species, age, and sex of the fish (if known)?
    - How long has the fish been kept, where did it come from, and what is the history of its arrival to the current system?
    - How long has the system been in operation and what is the water source?
    - How long has the client been keeping fish in general (gauge general knowledge base)?
    - What problem(s) has (have) been observed and for how long?
    - What other fish and other animals are in this same tank or system (species, number, sex, and ages (if known))?
    - Are any of these other fish showing any disease signs, and if so, what signs and for how long?
    - Any new introductions of fish, plants, substrate, decorations? Where did these come from?
    - Any previous problems in this or other tanks in the same household/facility?
    - What size is the tank, and what type of lighting and filters (biological, mechanical, other)?
    - What type of décor, other equipment, plants, and substrate are in the tank?
    - Any potential exposure to toxins (pest control, house cleaning/cleaning products, water changes without adequate dechlorination)?
    - What are general management protocols for tank and filter maintenance? Is there any schedule for water changes and water quality monitoring?
    - What are the results of your most recent water quality test (if done)?
    - What and how often are fish fed? Are the fish (sick and healthy) still eating normally?
    - Is tank mate aggression observed?
    - Has the fish changed color (an early sign of many different diseases) or exhibited any other behavioral changes or gross abnormalities?
    - What treatment(s) (water changes, drugs, doses) have been used so far, and when?

Water Quality/Chemistry/Management

Rule out water quality problems first; test the water (see "Aquarium Systems and Water Quality" section). Also rule out system and management issues.

FISH PHYSICAL EXAMINATION

Restraint
- Wear moistened, powder-free latex or nitrile gloves to avoid scale/epidermal damage.
- Keep fish on smooth, moist surface.
- One can restrain fish by placing in plastic bag with a little water, or use a soft net to assist and prevent fish from escaping.
- Anesthesia or sedation may be required.
- Cover eyes to calm fish, or dim lights, if possible.

Examples of abnormalities observable during physical examination (**Figure 5.1**):

a) Discus in head up position with erosion.
b) Angelfish with large ulceration.

Figure 5.1 General disease signs in tropical fish. A. This discus is thin, poorly colored, swimming in an abnormal position, and has a depigmented lesion above the lateral line on the right side. B. This angelfish has a raised, ulcerative lesion of the left epaxial area. C. This cichlid appears thin, its fins are clamped, and there are nodular white lesions on the dorsal fin. D. This striated frogfish (*Antennarius striatus*) is displaying extreme coelomic distension. RP Yanong

c) Jack Dempsey juvenile with white masses on the dorsal fin.
d) Striated frogfish with coelomic distension.

General Considerations

- Become familiar with "normal" behavior, including swimming, typical positioning and alignment in the water column, and appearance for common aquarium fish species. Consider visiting a local public aquarium or aquarium store and/or joining a local aquarium or pond society.
- When confronted with unfamiliar species, do your due diligence (consider closely related species, speak with trusted fish colleagues (veterinary, advanced fish keepers, aquarists), read reference books, examine trusted websites.
- Observe fish behavior, attitude/position in the water, respiratory rate (opercular/gill flap movements).
- Examine fish: first "at a distance" while it is in the water/tank, and then under restraint. From head to tail, look for any abnormalities, grossly visible lesions, including in or near the mouth, opercula (gill plate) and gill cavity, and assess the ophthalmic, neurologic, and dermatologic systems.
- Body condition should also be assessed, although potential disease or normal changes (e.g., reproductive readiness) will affect morphometrics (e.g., ascites vs. mature ovaries) and should be taken into account. Clark et al. (2018) provide some guidelines for body condition scoring in zebrafish as a template (see **Figure 5.2**). Examination from the dorso-ventral view often provides the most rapid useful information. Very thin fish (i.e., BCS 1 in **Figure 5.1**; see also **Figure 5.2**), will often have more skeletal processes apparent, especially around the skull with more "sunken in" musculature, and will have a "tadpole" appearance (i.e., very large head relative to the body).

DIAGNOSTICS

- Rule out parasites/bacteria/water mold/fungus/viral lesions (i.e., lymphocystis) that may be identified by wet mount (see **Figure 6.1, 6.2**, and Chapter 11.).
- If there is a problem in a multi-fish system, observe other fish as described above, and sample some of their external tissues if possible. Be thorough. Fish often have mixed infections.
- If fish is moribund and terminal, or if this is a population issue with multiple fish affected and the client allows, necropsy can be very informative (see "Necropsy" section in Chapter 8).

	Adult Zebrafish BCS	
	Lateral View	Dorsal View
BCS 1: • Head larger than body (big head) • Lateral- concave ventral surface between head and abdomen (narrow abdomen) • Dorsal- body is more narrow than head and linear • **Fish is thin (emaciated)**		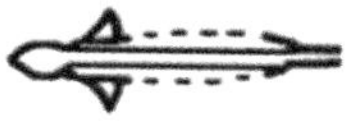
BCS 2: • Head and body equal size • Lateral- flat ventral surface between head and abdomen • Dorsal- head and width of abdomen are equal • **Fish is underconditioned**		
BCS 3: • Body larger than head • Lateral- slight convex ventral surface • Dorsal- head is slight smaller to a fusiform body • **Fish is well-conditioned**		
BCS 4: • Body significantly larger than head • Lateral- body moderately convex ventral surface • Lateral- Symmetrical ventral surface • Dorsal- head visually smaller to a moderately distended abdomen • **Fish is over-conditioned**		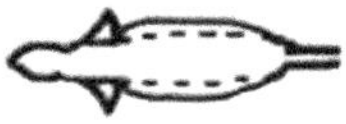
BCS 5: • Body significantly larger than head • Lateral- body significantly convex ventral surface • Lateral- Symmetrical or asymmetrical ventral surface • Dorsal- head visually smaller to a significantly distended abdomen • **Fish is obese (large)**		

Figure 5.2 This zebra danio (*Danio rerio*) body condition score (BCS) chart can be applied to other species of fishes. (Clark et al. 2018)

- Evaluation of feces: do not collect from bottom of tank/pond; opportunistic organisms may confuse a diagnosis. Isolate fish in a clean bucket or plastic bag and wait for fish to defecate. If necessary, place fish in anesthetic solution; fish will frequently defecate as it relaxes. Can also perform a saline rectal wash.

Endoscopic Evaluation
- A rigid endoscope can be used to examine the oral and branchial cavities for lesions and parasites. The magnification provided can, in some cases, diagnose even protistan disease.

EMERGENCY THERAPY/FIRST AID

- Take history; test water.
- Physical examination.
- Biopsy (skin, fin, gills) or necropsy.
- Take blood sample.
- Radiograph/Imaging.
- Fluid therapy for marine fish: use 20 ml/kg/day LRS ICe or PO, and/or consider dropping salinity in a hospital tank to 26–28 ppt.
- Freshwater fish don't drink (hyperosmotic) and don't usually need fluids.
  - It may be helpful to add 1–3 ppt (g/L) NaCl to a hospital tank for a freshwater fish to help with osmoregulation.
- Marine fish drink often and are hypoosmotic to seawater, so reducing salinity may help alleviate some osmotic stress.
- Dexamethasone injectable for shock 1–2 mg/kg IM or ICe.
- Epinephrine (for shock/cardiac arrest 0.2–0.5 ml IM or ICe of 1:1000).

Figure 5.3 Tube feeding a fish. Tube is inserted past the esophagus into the stomach (for fish with stomachs).

Tube Feeding
- Add water to regular fish feed to make a slurry, or obtain appropriate similar commercial fish feed (e.g., fish herbivore or omnivore) to make a slurry.
- Most bony fish (e.g., cichlids) possess a saccular stomach. Use small volumes for goldfish and koi (lack reservoir/true stomach).
- Use flexible, blunt-tipped feeding tube – can use lubricating jelly.
- Avoid branchial arches when inserting tube into pharynx.
- Tube should be long enough to reach the stomach (in fish with stomachs). See **Figure 5.3**.

CHAPTER 6
ANTE-MORTEM DIAGNOSTICS

EXTERNAL WET MOUNT BIOPSIES

- External wet mount biopsies require immediate microscopic examination (have slides ready with clean drops of water (isosmotic to tank/system water for external wet mounts) and laboratory analysis.
- Take small samples for fresh evaluation – squash prep of external (skin scrape, gill and fin biopsy) and internal tissues, for histology (formalin fixation), and, if warranted, for virology and molecular diagnostics (contact laboratory for preferred sample/preservation methods).
- See **Figures 6.1** and **6.2**.
- Skin: Use a coverslip to scrape over area of concern to collect mucus and epidermal tissue. If there are no obvious areas of concern, sample areas that receive less flow or turbulence and/or areas with softer, more exposed surfaces (e.g., behind pectoral or pelvic fins).
- Fin/Gill biopsy: Cut fin biopsy and 1–2 mm of gill tissue with fine scissors (+/- sedation for restraint). Place on slide with drops of water isosmotic (i.e., fresh, brackish, salt) to water in which fish is being kept for immediate analysis. Tank water can be used if it is clear. Some environmental problems may be suggested based on gill biopsy results. For example, telangectasia or "ballooning" of gill capillaries can be caused by elevated ammonia and water with excess particulates or organics can help initiate bacterial gill hyperplasia or heavy infestations of some parasites including trichodinids or sessile ciliates.
- See Chapter 11 tables and figures of parasites.

BACTERIAL CULTURE – NON-SACRIFICIAL

- For survival procedures, if the fish is large enough (>10 cm), a blood culture may be helpful to determine the presence of a systemic infection, and a culture of the leading edge of an ulcer may be helpful (although more challenging to interpret).

DOI: 10.1201/9781003057727-6

- For blood cultures, sedate the fish, then rinse the external surface around the area where the needle will be inserted with sterile saline prior to blood draw. Typically, the caudal vessels are easiest to sample (see **Figures 6.1, 6.2–6.6**). For best culture results, incubate a drop or two of blood in liquid bacterial growth media (such as brain–heart infusion) for

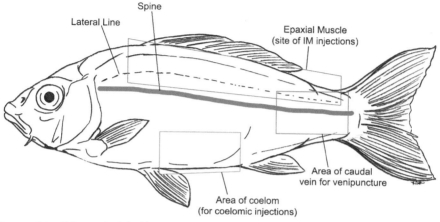

Figure 6.1 Schematic fish illustrating injection and blood collection sites. Drawing by Kirstin Cook

Figure 6.2 Photographs illustrating external biopsy collection. Clockwise from top left: skin scrape near pectoral and pelvic fins, skin scrape - dorsal, gill biopsy, fin clip. RP Yanong

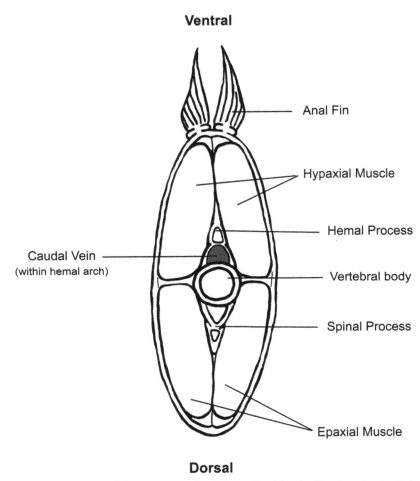

Figure 6.3 Schematic fish in cross section illustrating blood collection site (caudal vein/caudal vessels in hemal arch). Drawing by K Cook

24–48 hours at 28°C (warm water fish) (Klinger et al. 2003). Work with a microbiology lab familiar with fish isolates.
- Cultures from ulcers are more challenging due to presence of opportunists. The leading edge of an ulcer can be cultured after rinsing the surface well with sterile saline. Select a scale at the very edge of the ulcer, remove, and culture beneath it within the dermis/muscle interface.
- See "Necropsy" for post-mortem bacterial cultures.

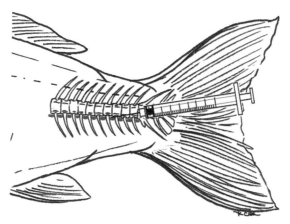

Figure 6.4A Schematic fish in lateral recumbency illustrating lateral blood collection approach (caudal vein in hemal arch). Drawing by K Cook

Figure 6.4B Photograph demonstrating caudal vessels lateral draw in an African cichlid. RP Yanong

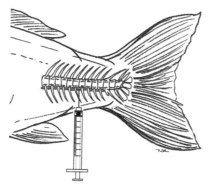

Figure 6.5 Schematic fish in lateral recumbency illustrating ventral blood collection approach (caudal vein in hemal arch). Drawing by K Cook

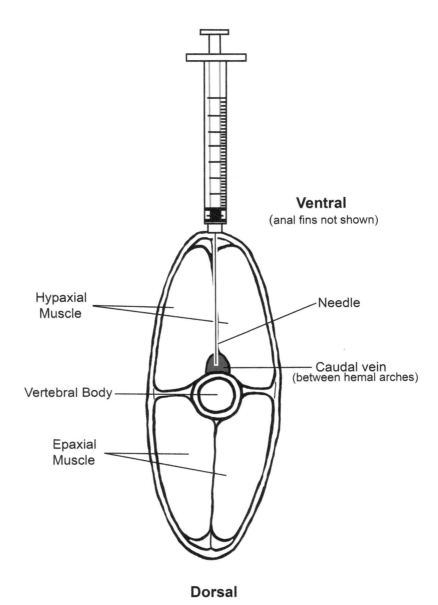

Figure 6.6 Schematic fish in dorsal recumbency illustrating ventral blood collection approach (caudal vein in hemal arch). Drawing by K Cook

BLOOD COLLECTION TECHNIQUES (SEE FIGURES 6.1–6.14)

- Become familiar with sites/approaches (see **Figures 6.1–6.14**, and–for common teleost and elasmobranch sites); depending upon the status of the animal, size, and anatomy, collection may not be straightforward or possible.
- The most common blood collection sites include caudal vessels (hemal arch), cardiac, gill arch, dorsal fin sinus (shark), wing (ray).
- Make blood smears and process samples for clinical pathology.
- Knowledge of presence, location, and size of hemal spines/vertebral processes is helpful, especially for caudal vessel access.
- Use heparinized syringe and needle, if possible, because fish blood clots rapidly.
- Needle gauge and length should be selected based on estimated vessel size and fish size as well as on approximate location of targeted vessels or organ (i.e., heart).
- Caudal vein/vessels.
  - Technique 1: For many fish, the lateral line is a good landmark for caudal vessel blood collection. With fish in lateral recumbency, visualize the location of the vessels within the hemal arch and direct the needle at approximately 40° toward the ventral margin of the thin lateral line caudal to the vent and cranial to the tail/caudal peduncle. Insert until resistance from the spinal column is felt, then "walk" or slowly angle ventrally until needle slips into the hemal space containing the vessels just ventral to the spine. If necessary, pull the needle out completely and reposition (see **Figures 6.4A and 6.4 B**).
  - Technique 2: Place fish in dorsal recumbency, use a midventral approach, and insert the needle at a point midway between the anal fin and tail. Insert until resistance from the vertebral body is felt, then withdraw needle slightly to hemal space containing caudal vessels (see **Figures 6.4–6.5**).
- Gill arch vessels.
  - Technique 3: This method is suggested for larger fish. Sedate and place fish in dorsal recumbency. Insert needle into the gill arch (central/midline of the arch) at an approximately 70°–80° degree angle. A slight resistance will be felt entering the arch prior to reaching the vessels (see **Figure 6.7**).
- Cardiac draw.
  - Technique 4: This method is suggested only for euthanized fish and requires approximate knowledge of the position of the heart in the species of interest to guide needle length and placement. To aid in

appropriate placement, immediately after euthanasia, place the fish in dorsal recumbency, and, for the African cichlid example (see **Figure 6.8**), the needle is placed ventral midline at the level of the edge of the operculum (gill plate) at an approximate 80°–85° degree angle toward the base of the pectoral fin. Slowly insert the needle while drawing gently back on the syringe until a flash is seen.
- Cut tail/peduncle draw.
  - Technique 5: This method is suggested for euthanized fish fewer than or equal to 3 inches (7.5 cm) in total length. Immediately after euthanasia, the caudal peduncle of the fish is cut with a sharp blade and a microhematocrit tube is used to collect blood from the exposed caudal vessels (see **Figure 6.9**).
- Lateral access to bulbus/cardiac draw.
  - Technique 6: This method is suggested for fish fewer than or equal to 3 inches (7.5 cm) in total length. Immediately after euthanasia, the right (or left) pectoral fin is cut. Metzenbaum scissors are then used to cut into the bulbus arteriosus/heart, and a microhematocrit tube is used to draw blood from the cut organ. This technique is used successfully in zebra danios and other smaller bodied fish (see **Figure 6.10**).
- Elasmobranch – wing vessel draw in rays.
  - Technique 7: Radial wing vessels (arterial/venous dependent on sample site) run throughout the wings of rays and can be accessed for blood collection as illustrated in **Figure 6.11**. Note: A larger vein in batoids – the mesopterygial vein – (see **Figure 6.12** for cross section schematic) can be accessed for blood collection as well as fluid administration, and readers may refer to Westmoreland et al., 2019 for more specific methods.
- Elasmobranch – tail vasculature blood draw.

Figure 6.7 Photograph demonstrating gill arch vessels blood draw, a method suggested for larger fish. RP Yanong

Blood Collection Techniques (See Figures 6.1–6.14)

Figure 6.8 Photograph demonstrating cardiac ventral draw in an African cichlid, a method for use in euthanized fish. RP Yanong

Figure 6.9 Photograph demonstrating cut tail/caudal peduncle draw in an African cichlid, a method for use in euthanized fish 3 or fewer inches in length. RP Yanong

Figure 6.10 Photograph demonstrating a lateral approach, bulbus arteriosus/cardiac draw from a small African cichlid, a method for use in euthanized fish. Immediately after euthanasia, the right pectoral fin is cut. Metzenbaum scissors are then used to cut into the bulbus arteriosus/cardiac region, and a microhematocrit tube is used to draw blood. This technique is used successfully in zebra danios and other smaller bodied fish. RP Yanong

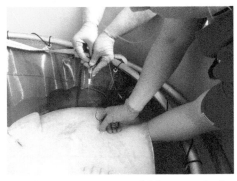

Figure 6.11 Photograph demonstrating wing phlebotomy via the mesopterygial vein in a sting ray. E Christiansen

- Technique 8: Blood can be drawn from the tail of elasmobranchs in a manner similar to that of other vertebrates (see **Figures 6.13** and **6.14**). With the animal in dorsal recumbency, a needle is inserted at a 90° angle to access the vessels within the hemal arch, located ventral to the spinal column.
- Elasmobranch –dorsal sinus blood draw.

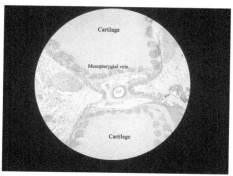

Figure 6.12   Histological micrograph demonstrating mesopterygial vein in section. CA Harms

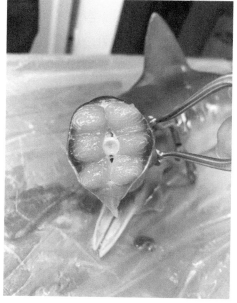

Figure 6.13   Photograph of sectioned shark tail showing target tail vasculature ventral to spinal column. A. Delaune, Mississippi Aquarium

Figure 6.14 Photograph demonstrating blood collection from shark tail vasculature, with shark in dorsal recumbency. A. Delaune MS Aquarium

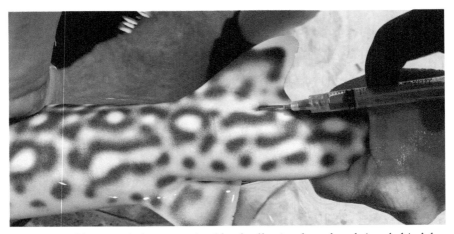

Figure 6.15 Photograph demonstrating blood collection from dorsal sinus behind dorsal fin. Alexa Delaune MS Aquarium

- Technique 9: An alternative blood collection site for sharks is the dorsal sinus, located behind/adjacent to the dorsal fin of sharks (see **Figure 6.15**). The needle is angled slightly above parallel in an area caudal to the dorsal fin. Studies have demonstrated that blood analyses from this site may differ from other more appropriate venipuncture sites.

DIAGNOSTIC EVALUATION OF SAMPLES

- Visualize parasites, fungi/water molds, bacteria, other pathogen-related lesions (e.g., lymphocystis) from skin, fin, or gill biopsies (see Chapter 11).
- Gram stain blood (blood smear may reveal sepsis) and brain or kidney impression smears (after heavy blotting to remove excess blood from these tissues). (Note: Gram-negative bacteria at very low levels may be normal on skin, gills, water).
- Obtain a culture from blood, kidney, or brain for a bacterial diagnosis.
- Preserve tissue samples in 10% neutral buffered formalin for laboratory analysis (use a laboratory familiar with fish diseases and microbiology).

HEMATOLOGY

- Higher water temperature, stress, and disease affect hematologic values.
- Manual analysis methods (hemocytometer, refractometer, centrifuge) are required due to cell size and nucleation (red blood cell (RBCs) and thrombocytes) that skew automated counts.
- Thrombocytes are the most abundant non-erythroid blood cell in carp and goldfish.
- RBCs are oval, nucleated, and larger than mammalian RBCs. Packed cell volume (PCV) lower than 20% indicates anemia.
- White blood cells (WBCs): Slide estimated total WBC counts (using the avian WBC slide estimate method, which relies on ratios of RBCs to WBCs) are lower than values obtained using a hemocytometer. Pacu RBCs (and likely other fishes, too) are present at 50%–90% of the concentration observed in birds, so the avian method should be adjusted ([Mean WBC/10 HPF x 2000] + 6300).
- Reference ranges for selected species are contained in **Tables 6.1–6.2**.

Table 6.1 Hematology Reference Ranges For Selected Fishes*

PARAMETER	GOLDFISH (Groff and Zinkl, 1999; Adamovicz et al., 2017)	KOI (Palmeiro et al., 2007; Tripathi et al., 2004	RED PACU* (Tocidlowski et al., 1997)	TILAPIA (Hrubec ete al., 2000)	SANDBAR SHARK* (Arnold, 2005)
PCV (%)	26 +/- 1 31+/-7.3	35 (24–43)	26 (22–32)	33 (27–37)	22.8 (21–23)
RBC (106/µl)	1.5 +/- 0.1	1.67 +/- 0.08	1.7 (1.2–2.9)	–	–
Hgb (g/dl)	9.1 +/- 0.4	6.9 +/- 1.6	–	2.31 (1.91–2.83)	–
MCV (fl)	–	178.2 +/- 31.7	–	135.7 (115–183)	–
MCH (pg)	–	40.2 +/- 6.5	–	34.9 (28.3–42.3)	–
MCHC (g/dl)	–	21.6 +/- 3.3	–	25.7 (22–29)	–
WBC (103/µl)	–	24 +/- 5.6	33.5 (13.6–52.3)	75.7 (21.6–154.7)	–
Heterophils (%)	29 +/- 3	–	5.2 (0.3–36.7)	–	5.2 (0.3–36.7)
Lymphocytes (%)	70 +/- 5	79.1 +/- 12.4	84 (53–96)	–	50 (40–55)
Monocytes (%)	1 +/- 0.1	2.9 +/- 1.3	4 (0.8–11.2)	2.0	3 (2–6)
Eosinophils (%)	–	–	0.3 (0.3–0.7)	0.44	0.3 (0.3–0.7)
Basophils (%)	–	4.6 +/- 2.6	–	–	–
Neutrophils (%)	–	10.9 +/- 7.7	–	2.38	46 (40–58)
Small lymphocytes (%)	–	–	–	80.8	–
Large lymphocytes (%)	–	–	–	14.2	–
Thrombocytes (103/µl)	–	–	–	52.7 (25.1–85.2)	–

Table 6.1 (Continued)

PARAMETER	LIONFISH (Anderson et al., 2010)	CLOWNFISH (AMPHIPRION SPP.) (Wright et al., 2021)	LARGEMOUTH BASS (WILD) (Whitehead et al., 2019)	PINFISH (Collins et al., 2016)	SPOTTED EAGLE RAY (Greene et al., 2018)
PCV (%)	34 (27–44)	21	44.9 +/− 5.4	34.4 +/− 9.5	28 +/− 3
RBC (106/µl)	–	–	194.5 +/− 37.4	–	–
Hgb (g/dl)	–	–	–	11.7 +/− 3.2	7.31 +/− 1.09
MCV (fl)	–	–	–	–	–
MCH (pg)	–	–	–	–	–
MCHC (g/dl)	–	–	–	–	–
WBC (103/µl)	4.0 (2.0–8.2)	4.3	7.08 +/− 1.86	25.8 +/− 14.4	18 +/− 9
Heterophils (%)	45.5 (13–72)	–	–	–	*25.65 +/− 17.58
Lymphocytes (%)	21.5 (7–67)	70	67.7 +/− 13	80.6 +/− 16	71 +/− 19.1
Monocytes (%)	27.5 (16–51)	8	8.3 +/− 2.9	3.8 +/− 2.5	1.88 +/− 1.93
Eosinophils (%)	–	0	–	–	–
Basophils (%)	–	0	–	–	1.24 +/− 1.2
Neutrophils (%)	–	18	24.1 +/− 12.7	17.1 +/− 14.9	–
Small lymphocytes (%)	–	–	–	–	–
Large lymphocytes (%)	–	–	–	–	–
Thrombocytes (103/µl)	–	–	15.3 +/− 5.8	–	–

Table 6.1 (Continued)

PARAMETER	FRESHWATER STINGRAY (POTAMOTRYGON SPP.)++ (Brito et al., 2015)
PCV (%)	21.5 +/- 2.7
RBC (106/µl)	0.97 +/- 0.175
Hgb (g/dl)	4.67 +/- 0.62
MCV (fl)	263 +/- 43.5
MCH (pg)	52.6 +/- 8.6
MCHC (g/dl)	22.1 +/- 2.3
WBC (103µl)	2.75 +/- 0.48
Heterophils (%)	18.2
Lymphocytes (%)	25.8
Monocytes (%)	7.2
Basophils (%)	3.7
Neutrophils (%)	54.6
Small lymphocytes (%)	–
Large lymphocytes (%)	–
Thrombocytes (103/µl)	1.32

PCV (%): Striped bass = 42 (34–48); Southern stingray* = 22 (15–25); Bonnethead shark* = 24 (17–28)
* Combination of fine and coarse eosinophilic granulocytes (Greene et al., 2018)
++Average of four species (*P. motoro, P. falkneri, P. orbignyi, P. scobina*) (Brito et al., 2015)

Table 6.2 Blood Chemistry Reference Ranges for Fish

PARAMETER	GOLDFISH (Groff and Zinkl, 1999; Adamovicz et al., 2017)	KOI* (Palmeiro et al., 2007; Tripathi et al., 2004)	STRIPED BASS (Hrubec et al., 2000; Noga et al., 1999)	LIONFISH (Anderson et al., 2010)	CLOWNFISH (AMPHIPRION SPP.) (Wright et al., 2021)	PINFISH (LAGODON RHOMBOIDES) (Collins et al., 2016)	SPOTTED EAGLE RAY (AETOBATUS NARINARI)+ (Greene et al., 2018)
ALP (IU/L)	–	–	–	35 (16–66)	–	–	27.1 +/– 7.7
ALT (IU/L)	106 +/– 9	–	–	1.0 (1.0–7.0)	–	–	16.3 +/– 0.96
Anion gap (mmol/L)	–	17 (14–23)	24 +/– 1	–	–	–	–
Bicarbonate (mmol/L)	–	6 (3–8)	–	–	–	–	–
AST (IU/L)	229 (111–43)	121 (40–381)	45 +/– 21	69 (24–236)	–	–	8.42 +/– 4.5
Calcium (mg/dL)	9.87 (6.2–13.5)	8.7 (7.8–11.4)	11.1 +/– 0.2	10.8 (9.4–28.4)	5.2 (1.2–10)	6.24 +/– 0.72	17.5 +/– 0.81
Chloride	–	114 (108–119)	144 +/– 2	149 (142–162)	140	–	277 +/– 17.4
Cholesterol (mg/dL)	–	–	–	159 (75–252)	–	–	102.4 +/– 16.9
Triglycerides (mg/dL)	–	–	–	298 (59–661)	–	–	74 +/– 37
Amylase (U/L)	–	–	–	1 (1–2)	–	–	–
Lipase (U/L)	–	–	–	8 (2–32)	–	–	–
Cortisol (µg/dl)	–	23.7	–	–	–	–	–
Creatine kinase (IU/L)	5319.6 (853–8877)	4123 (80–9014)	–	860 (198–4372)	–	–	102.75 +/– 77.4
Creatinine (mg/dL)	–	–	0.3 +/– 0.0	0.1 (0.1–0.2)	2.0	–	–
Glucose (mg/dL)	73 +/– 9	37 (22–65)	118 +/– 10	26.5 (10–49)	45 (20–79)	170 +/– 101	40.3 +/– 11.6
LDH (IU/L)	–	359 (41–1675)	164 +/– 54	–	–	–	–
Magnesium (mEq/L)	–	–	–	3.0 (2.1–4.5)	–	–	2.97 +/– 0.295

Table 6.2 (Continued)

PARAMETER	GOLDFISH (Groff and Zinkl, 1999; Adamovicz et al., 2017)	KOI* (Palmeiro et al., 2007; Tripathi et al., 2004)	STRIPED BASS (Hrubec et al., 2000; Noga et al., 1999)	LIONFISH (Anderson et al., 2010)	CLOWNFISH (AMPHIPRION SPP.) (Wright et al., 2021)	PINFISH (LAGODON RHOMBOIDES) (Collins et al., 2016)	SPOTTED EAGLE RAY (AETOBATUS NARINARI)+ (Greene et al., 2018)
Osmolality (mOsm/kg)	–	–	356 +/– 2	–	–	–	–
Phosphorus (mg/dL)	9.57 (5.1–16.3)	6.1 (3.5–7.7)	9.8 +/– 0.2	10.7 (7.9–20.8)	–	–	3.88 +/– 0.74
Potassium (mEq/L)	1.87 (0.1–5.6)	3.4 (2.7–4.3)	3.3 +/– 0.2	2.9 (1.9–4.0)	4.8 (3.1–8.1)	4.7 +/–1.4	4.36 +/– 0.47
Protein, total (g/dL)	2.99 (1.2–5.0)	2.0 (1.4–2.7)	4.6 +/– 0.1	4.0 (2.2–6.3)	3.5 (2.1–5.0)	7.2 +/–1.1	5.72 +/– 0.44
Albumin (g/dL)	2.13 (0.5–3.2)	0.9 (0.6–1.1)	1.3 +/– 0.0	1.0 (0.6–2.0)	–	–	–
Globulin (g/dL)	0.7 (0.3–1.2)	1.1 (0.8–1.6)	–	2.9 (2.2–4.5)	–	–	–
A:G (ratio)	4.3	–	0.4 +/– 0.0	0.3 (0.2–0.5)	–	–	–
Sodium (mEq/L)	133.8 (126–140)	134 (112–141)	174 +/– 2	172 (168–177)	163 (141–180)	169+/– 7	297.25 +/– 17.44
Total CO2 (mmol/L)	–	–	10.7 +/– 0.9	–	–	8.94 +/–2.84	–
pH	–	–	–	–	–	7.0 +/–0.28	–
Lactate (mmol/L)	–	–	–	–	–	9.93 +/– 4.56	–
Urea nitrogen (mmol/L)	10 +/– 0	–	–	3.0 (2.0–5.0)	1.1 (1.0–1.8)	–	397.3 +/– 39.7

Table 6.2 (Continued)

PARAMETER	RED PACU (Sakamoto et al., 2001)	TILAPIA* (Hrubec et al., 2000)	FRESHWATER STINGRAY (POTAMOTRYGON SPP.)++ (Brito et al., 2015)
AP (IU/L)	–	22 (15–39)	87.25 (32–124)–
ALT (IU/L)	–	–	18.75 (13–25)
GGT	–	–	8.8 (8.0–9.9)
Anion gap (mmol/L)	6.9 (1.2–12.5)	–	–
Bicarbonate (mmol/L)	49 (0–125)	–	–
AST (IU/L)	–	26 (9–102)	94.5 (68–119)
Calcium (mg/dL)	10.8 (9.5–12.5)	31 (13.6–69.4)*	9.2 (8.1–9.7)
Chloride	138 (146–159)	136 (128–142)	156 (151–159)
Cholesterol (mg/dL)	–	189 (110–318)	54.8 (21.8–88.7)
HDL-C (mg/dL)	–	–	8.8 (8.1–9.7)
Triglycerides (mg/dL)	–	–	48.8 (37.5–66.7)
Cortisol (µg/dl)	–	–	–
Creatine kinase (IU/L)	–	–	13,415 (904–7248)
Creatinine (mg/dL)	0.3 (0.2–0.4)	0.2 (0.1–0.5)	0.35 (0.3–0.4)
Glucose (mg/dL)	–	46 (30–69)	53 (39–67)
LDH (IU/L)	238 (65–692)	–	–
Magnesium (mEq/L)	–	2.5 (1.9–3.5)	–
Osmolality (mOsm/kg)	–	–	–

Table 6.2 (Continued)

PARAMETER	RED PACU (Sakamoto et al., 2001)	TILAPIA* (Hrubec et al., 2000)	FRESHWATER STINGRAY (POTAMOTRYGON SPP.)++ (Brito et al., 2015)
Phosphorus (mg/dL)	7.3 (4.1–8.9)	9.1 (5.5–22.1)*	8.56 (8.1–9.1)
Potassium (mEq/L)	3.9 (2.7–5.0)	4.3 (3.5–5.4)	9.8 (6.0–14.9)
Protein, total (g/dL)	–	3.0 (2.9–6.6)*	2.5 (2.0–3.0)
Albumin (g/dL)	0.9 (0.5–1.0)	1.8 (1.3–2.6)	0.275 (0.1–0.4)
Globulin (g/dL)	–	2.1 (1.6–4.2)	1.85 (1.5–2.2)
A:G (ratio)	–	–	0.15
Sodium (mEq/L)	150 (146–159)	151 (139–160)	167 (156–175)
Total CO2 (mmol/L)	7.5 (6–10)	–	–
Urea nitrogen (mmol/L)	–	–	30.4 (29–38.3)
Uric acid mg/dL	–	–	0.975 (0.5–1.3)
T3 (ng/dL)	–	–	1.9 (1.6–2.0)
T4 (ng/dL)	–	–	2.2 (1.1–2.5)

Table 6.2 (Continued)

PARAMETER	SOUTHERN STINGRAY* (Cain et al., 2004)	BONNETHEAD SHARK* (Harms et al., 2002)
AP (IU/L)	—	—
ALT (IU/L)	—	—
Anion gap	—	-5.8 (-15.7 to -7.5)
Bicarbonate (mmol/L)	—	3 (0–5)
AST (IU/L)	14.5 (3.6–61.2)	42 (15–132)
Calcium (mg/dL)	16.5 (12.1–19.3)	16.8 (15.8–18.2)
Chloride	342 (301–362)	290 (277–304)
Cholesterol (mg/dL)	—	—
Creatine kinase (IU/L)	80.5 (11.7–296.5)	82 (18–725)
Creatinine (mg/dL)	---	—
Glucose (mg/dL)	30.5 (16.9–42.4)	184 (155–218)
Lactate (mmol/L)	3.1 (<2.0–6.2)	<5 (<5–11)
LDH (IU/L)	—	—
Magnesium (mEq/L)	—	—
Osmolality (mOsm/kg)	1065 (1008–1144)	1094 (1056–1139)
Phosphorus (mg/dL)	4.7 (3.0–6.4)	8.8 (5.9–12.7)
Potassium (mEq/L)	5.0 (3.2–6.4)	7.3 (5.7–9.2)
Protein, total (g/dL)	2.6 (1.6–3.2)	2.9 (2.2–4.3)

Table 6.2 (Continued)

PARAMETER	SOUTHERN STINGRAY* (Cain et al., 2004)	BONNETHEAD SHARK* (Harms et al., 2002)
Total solids (g/dL)	5.6 (4.2–6.0)	—
Albumin (g/dL)	—	0.4 (0.3–0.5)
Globulin (g/dL)	—	2.6 (1.9–3.8)
A:G (ratio)	—	0.1 (0.1–0.2)
Sodium (mEq/L)	315 (296–326)	282 (273–292)
Total CO2 (mmol/L)	—	—
Urea nitrogen (mmol/L)	444 (423–462)	1004 (944–1068)

*Values listed are means except for the red pacu hematology, koi chemistry elasmobranch data, and lionfish data that are medians. The ranges listed for the southern stingray are 10th/90th percentiles. In some cases, the data is not based on a large sample size. These values are only meant to be guidelines. Age of fish, time of year, and water temperature may all affect "normal" clinical pathological data.

* Levels lower for tilapia raised in a low-density system.

+Values averaged from two different labs as part of the study (Greene et al., 2018).

++Average of four species (*P. motoro, P. falkneri, P. orbignyi, P. scobina*) (Brito et al., 2015).

*Values listed are means except for the red pacu hematology and koi chemistry and elasmobranch data, which are medians. The ranges listed for the southern stingray are 10th/90th percentiles. In some cases, the data is not based on a large sample size. These values are only meant to be guidelines. Age of fish, time of year, and water temperature may all affect "normal" clinical pathological data.

CHAPTER 7
IMAGING

RADIOGRAPHY AND ULTRASONOGRAPHY: GENERAL COMMENTS

- Use to diagnose swim bladder disorders, gastrointestinal (GI) impactions or gas, foreign bodies, as well as liver, spleen, and skeletal abnormalities.
- Can use contrast media, as with other taxa, to assess GI abnormalities.
- Take radiographs of normal fish to establish technique and obtain normal views.
- May be done without anesthesia on fish in lateral recumbency on plate protected with a plastic bag. Some fish can be placed directly on the protected plate, others can be restrained by placing in a plastic bag with a little water.
- Large catfish, koi, and cichlids may require sedation.
- Dim room lights to reduce photophobia.
- Obtain dorsoventral (DV) or ventrodorsal (VD) views by rotating radiograph machine for a horizontal beam.
- Grid is unnecessary; high-detail intensifying screens and dental film are recommended.
- If normal fish can be used as a reference, in addition to organ morphology, placement, abnormalities, and ultrasound may also help with assessment of gonadal development.
- See **Figures 7.1-7.6**.

DOI: 10.1201/9781003057727-7

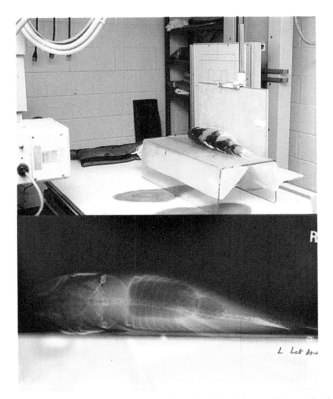

Figure 7.1   An anesthetized koi (*Cyprinus rubrofuscus*) is being radiographed with a horizontal beam lateral dorsal approach. The resulting radiograph (below) shows the cranial swim bladder compartment deviated slightly to the right. GA Lewbart

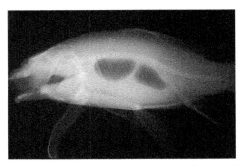

Figure 7.2   An anesthetized butterfly koi (*Cyprinus rubrofuscus*) radiograph in lateral recumbency. Both swim bladder compartments are clearly visible. GA Lewbart

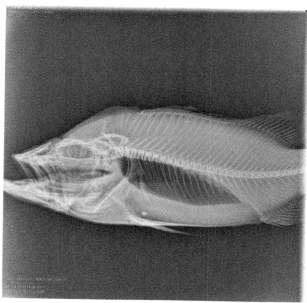

Figure 7.3 A peacock bass (*Cichla* sp.) in lateral recumbency illustrates a physoclistous fish with a single swim bladder compartment. There is also a radio dense object, possibly a small stone, in the gastrointestinal tract. S Boylan

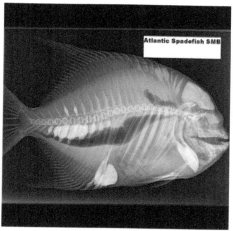

Figure 7.4 A spadefish (*Chaetodipterus faber*) in lateral recumbency. This species is very interesting for its dorsal swim bladder compartment and its heavily ossified bones. S Boylan

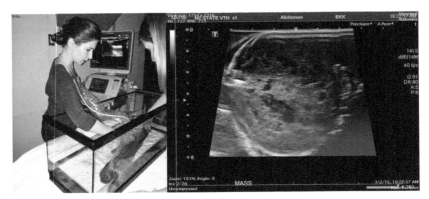

Figure 7.5   Ultrasonography can be performed with the fish in or out of the water. This bowfin (*Amia calva*) from a large display aquarium had an abdominal wall melanophoroma that was successfully removed via surgery. GA Lewbart

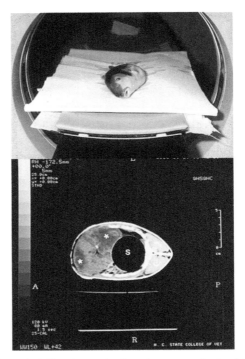

Figure 7.6   This showa koi (*Cyprinus rubrofuscus*) was anesthetized for about 6 or 7 minutes for this computed tomography (CT) procedure in 1996. Newer machines are very fast, and a 1 kg fish can be imaged in about 90 seconds. The patient had an undifferentiated abdominal sarcoma, mostly likely testicular in origin, that was successfully removed via surgery (this fish is also pictured in **Figure 7.1**). GA Lewbart

CHAPTER 8
POST-MORTEM EVALUATION

GENERAL COMMENTS ON SPECIMEN QUALITY

Ideal samples are clinically diseased, live fish that can be euthanized and necropsied immediately. Fish found dead may be presented for evaluation. Assessment of the freshness of the sample and knowledge of general necropsy techniques will help guide the approach to evaluation of (and/or rejection of) any dead samples provided. There are a number of veterinary pathology laboratories that will accept fish samples (including private, state, and university facilities), but the additional transit time and resulting autolysis should be weighed against a "fresher" in-house evaluation. Discuss the sample condition and potential rule-outs as well as shipping protocols with the laboratory personnel. If a sample is shipped out, one suggestion is to take skin, fin, and gill biopsies for in-house evaluation prior to processing for shipment.

The degree of freshness or autolysis will determine whether diagnostic value is high (if fresh dead, i.e., found moribund and died immediately prior to your arrival) or low (if dead overnight). Some suggested markers to evaluate freshness include:

1. Whether eyes are clear (suggesting fresher) or cloudy (less fresh – although this may be confounded if clinical signs included ocular opacity).
2. Color and integrity of the gills (red gills suggested freshness and pale pink and friable gills suggest autolysis/post-mortem degradation) – although some diseases may result in anemia or gill necrosis.
3. Smell – both externally and internally. A "rotten fish" odor externally or internally, with poor intracoelomic organ distinction strongly suggest significant post-mortem autolysis.

Information for carrying out an in-house necropsy is provided below.

DOI: 10.1201/9781003057727-8

NECROPSY SUPPLIES

- Compound microscope.
- Necropsy report form.
- Gloves (powder-free latex, nitrile).
- Necropsy tray/table.
- Dechlorinated freshwater or saltwater.
- MS-222 (tricaine methanesulfonate).
- Sodium bicarbonate or sodium carbonate (to buffer MS-222).
- Scissors (ophthalmic, medium and/or large with blunt and pointed tips).
- Scalpel (various blades/handles).
- Forceps (rat tooth, microdressing).
- Slides, cover slips.
- Paper towels.
- Needles/Syringes.
- Ruler.
- Scale.
- For some species: drill, Dremel, saw.

MICROBIOLOGY SUPPLIES

- Culture swabs/culturettes.
- Sterile loops, gauze.
- Burner (alcohol, gas).
- Alcohol (70% isopropyl alcohol/alcohol swabs (for fish disinfection prior to bacterial sampling) and 70%–80% ethanol (for burner/flame disinfection of instruments)).
- Matches/lighter.
- Glass jar/container to hold alcohol for disinfection of instruments.

Dead fish autolyze rapidly. Instruct the client to remove the fish (preferably when moribund and not dead) and bring it in as soon as possible. For fresh-dead fish (client either witnessed death or, less ideal, knows fish has been dead for <30min), have the client place it in a plastic bag (without added water) and refrigerate it.

- Bring a) system water and b) source water samples in two separate containers for evaluation. Aerated source water provides a comparative baseline.

- For moribund fish, prior to euthanasia, observe fish in a bag, bucket, or tank (ideally clear container) and document abnormal behavior, swimming, and lesions and note respiratory rate and effort.

WET MOUNT EVALUATION

- With practice, external biopsies for wet mount evaluations can be done rapidly on most live fish with minimal stress. If uncertain about technique, client, or fish reaction, and for more detailed external evaluation, sedate the fish with buffered MS-222 prior to biopsy and evaluation.
- Note: some data suggest that use of MS-222 may reduce parasite loads, especially if unbuffered.
- For wet mounts from external biopsies, add one to three drops of water spaced evenly (for skin, fin, gill) on a slide.
  - Use dechlorinated freshwater or spring water (not distilled) for freshwater fish.
  - Use full strength seawater for marine fish or isosmotic water for brackish water fish (*note: the slide sample may dry rapidly and form salt crystals, so evaluate immediately after prep).
- See **Figures 8.1–8.7**.

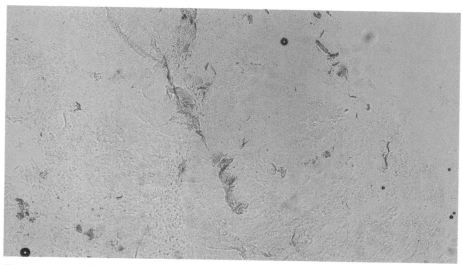

Figure 8.1A   Microscopic view of skin scraping with normal appearing mucus. RP Yanong

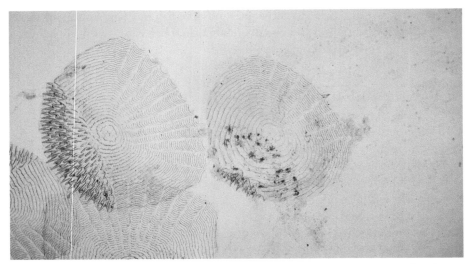

Figure 8.1B Microscopic view of skin scraping with normal appearing ctenoid scales. RP Yanong

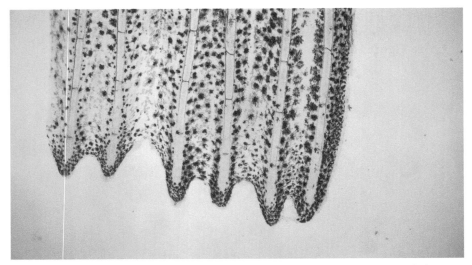

Figure 8.2 Normal microscopic view of fin tissue with healthy bony rays, epithelium, melanophores, and chromatophores. RP Yanong

- When sampling external tissues, only a small amount is needed. If the sample is so thick that after "squashing" the specimen, light cannot penetrate well through the sample, it will be too dark and difficult to evaluate. When observed during gross evaluation, target abnormal areas/lesions to sample for wet mounts (see **Figures 8.1–8.3**).

Figure 8.3 Normal microscopic view of gill tissue. Note long "finger-like" primary lamellae (supported by gill cartilage) comprised of numerous "plate-like" secondary lamellae that overlap each other and run the length of the primary lamellae in this micrograph. RP Yanong

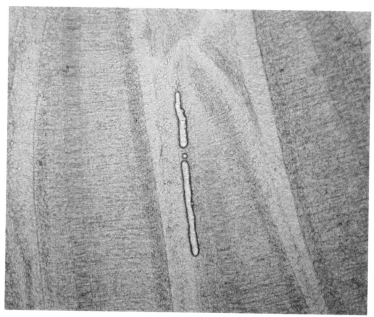

Figure 8.4 Gas bubbles within a branchial (gill lamellar) vessel. RP Yanong

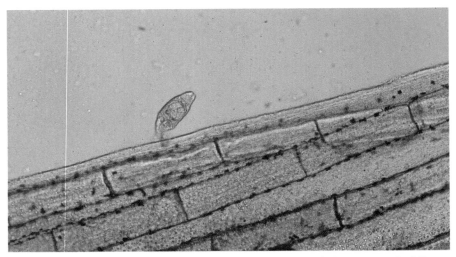

Figure 8.5 Microscopic view of a monogenean parasite on a fin biopsy sample. RP Yanong

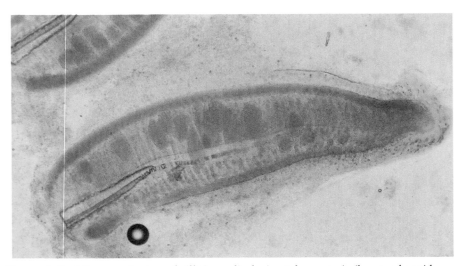

Figure 8.6 Microscopic view of gill tissue displaying telangectasia (large red ovoid structures). RP Yanong

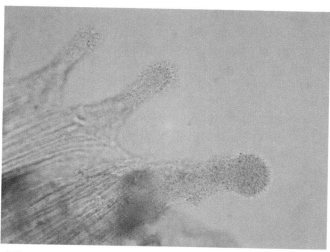

Figure 8.7   Microscopic view of *Flavobacterium columnare* on the fin. Note "haystack-like" formations that are often seen in wet mounts from affected fish. RP Yanong

- Place each tissue sample on a drop of water and then cover and squash the tissue with coverslip.
- Check for lesions/parasites microscopically (see **Figures 8.4–8.7**).

NECROPSY AND MICROBIOLOGY

Necropsy

- After wet mount external evaluation, euthanize (for information on methods, see Chapter 9) with tricaine methanesulfonate (MS-222; 500–1000 mg/L) buffered with sodium bicarbonate (MS-222: buffer →1:1 or 1:2) or sodium carbonate (0.5:1 or 1:1) and wait for 30 minutes after loss of buoyancy control and cessation of opercular movement/respiration as a one-step procedure. As a two-step procedure, immediately after loss of buoyancy control and cessation of opercular movement, cut at base of cervical spine and pith, or just pith.
  - Goldfish, koi, and cichlids may require pithing or another secondary method if MS-222 is used; if cultures are being taken for microbiology, the aseptic brain culture sampling can also serve as pithing.
- Gross physical examination (head to caudal fin/tail): observe all external structures, oral cavity and gill chambers, lesions, grossly visible parasites. Take photos to document the fish prior to and during dissection.

Microbiological Evaluation
- Have supplies and equipment ready.
- If you are submitting culturettes or aseptically sampled tissues, work with a laboratory familiar with fish microbiology and culture methods.
- Most common bacteria grow on tryptic soy agar with 5% sheep's blood (TSA + 5SB; i.e., blood agar).
  - Use more specific media for some (Ordal's/selective cytophaga agar for flavobacteria, Lowenstein-Jensen for mycobacteria, etc.).
- Aseptic technique is critical.
  - Use 70% isopropyl alcohol to disinfect outer surfaces that will be cut prior to microbiological sample collection; be sure to let surface dry or "flame" prior to incisions to avoid accidental target tissue exposure to alcohol.
- Primary target tissues (see **Figure 8.8**):
  - Brain and posterior kidney.
  - Liver, spleen, other structures as indicated.
  - Skin ulcers (leading edge), gonad, swim bladder, etc.
- Brain culture (see **Figure 8.9**):
  - After disinfection, allow alcohol to air dry.
  - With a sharp scalpel, slice off outer portion of skin/muscle/skull overlying brain.
  - In larger specimens, heavy duty instruments may be necessary or may require the use of a disinfected drill bit (and select different specimen for histo sample of the brain).

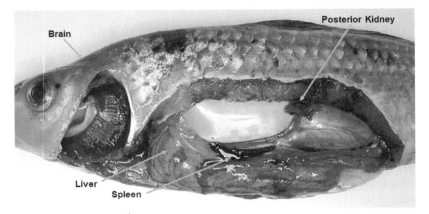

Figure 8.8   This annotated photograph indicates the most common and important sites for obtaining bacterial cultures. RP Yanong

Figure 8.9 This annotated photograph shows the exposed, fresh brain of a fish undergoing a necropsy. RP Yanong

- The brain in small to moderate size specimens should be easily exposed and can be cultured.
- Two approaches for kidney (see **Figure 8.10**):
- Dorsal approach:
  - Remove/cut dorsal fin and disinfect dorsal and dorso-lateral portion of body ventral (past areas overlaying the spine and swim bladder).
  - With sharp scissors, cut through the dorsal portion midline just past the spine ventrally (but not beyond the swim bladder to avoid cutting into the GI tract).
  - Ventral/lateral approach:
  - For most fish, position the fish in right lateral recumbency (lying with right side of body on surface).
  - Disinfect areas of body that will be cut to expose internal organs, swim bladder, and that overlies the posterior kidney.
  - Prior to disinfection, cut any fins that may overlap areas and cause contamination.
  - Palpate to identify coelom/epaxial musculature junction.
  - Incise immediately ventral to juncture to expose the swim bladder.
  - Cut the swim bladder dorsally to expose and sample the posterior kidney.
  - *Note: ulcers or other external lesions can be cultured, but results will be compromised by secondary and environmental bacteria.
  - For ulcers, disinfect around the leading edge with 70% isopropyl alcohol and/or rinse and wipe leading edge generously with sterile saline. Afterward, remove the scale at leading edge and culture under the scale (within the scale pocket) and into the musculature. Take several samples from different areas of the leading edge.

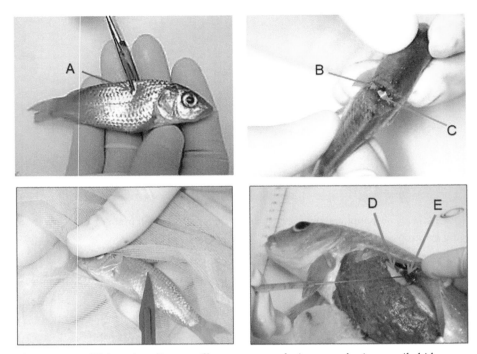

Figure 8.10 This series of images illustrate two techniques to obtain a sterile kidney culture or biopsy from a goldfish (*Carassius auratus*). Top row, first cutting through the epaxial musculature (A) and then spine; exposing the caudal kidney (B) and swim bladder (C). Bottom row, using a sterile scalpel blade to make a small slit into the coelom that can be used to initiate a window exposure to the body cavity and/or the swim bladder (D) and caudal kidney (E). RP Yanong

- Be sure to disinfect the entire body around the coelomic incision sites for internal evaluation (see next section) prior to wet mount evaluations of internal tissues, in case obtaining cultures of internal sites is necessary. After the coelom has been exposed, culture the spleen, liver, or other noted abnormal tissues.

INTERNAL EVALUATION

- After euthanasia (see Chapter 9), microbiological evaluation of internal organs should be completed (See "Microbiological Evaluation" section) as part of necropsy (see **Figures 8.11–8.20**). Prepare for internal tissue wet mount evaluation.

INTERNAL EVALUATION

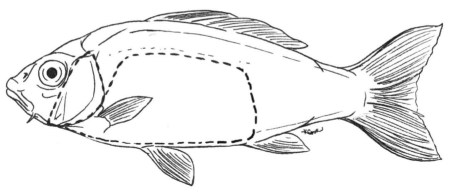

Figure 8.11 Schematic drawing of where to make cuts to remove the operculum and body wall of a koi (*Cyprinus rubrofuscus*). Drawing by K Cook

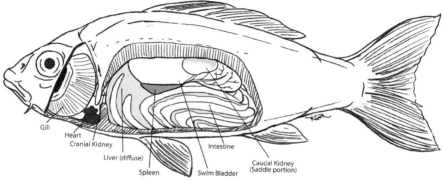

Figure 8.12 Schematic drawing of a koi (*Cyprinus rubrofuscus*) dissection with all of the major organs labeled after the body wall and operculum were removed. Cranial kidney may be more dorsally located. Drawing by K Cook

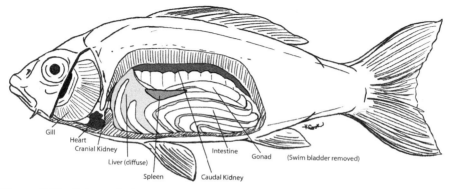

Figure 8.13 Schematic drawing of a koi (*Cyprinus rubrofuscus*) dissection with all of the major organs labeled after the swim bladder was removed. Cranial kidney may be more dorsally located Drawing by K Cook

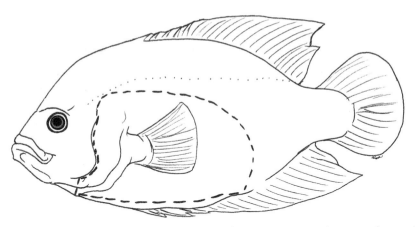

Figure 8.14 Schematic drawing of where to make cuts to remove the operculum and body wall of an oscar (*Astronotus ocellatus*). Drawing by K Cook

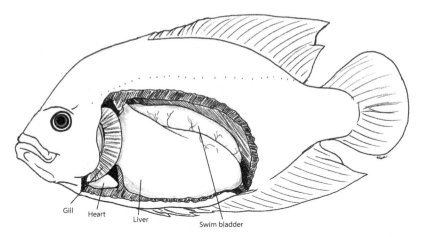

Figure 8.15 Schematic drawing of an oscar (*Astronotus ocellatus*) dissection with some of the major organs labeled. Drawing by K Cook

- Use dechlorinated (and not distilled) freshwater or physiological saline, and prep each slide with one to three drops of water (for each small tissue sample).
  - Remove operculum to expose and evaluate gills and then, after disinfection, remove body wall to expose internal organs.
  - Heart is anatomically separate from the coelom and is usually located posteriorly and ventral to the gills.
  - Take internal tissue samples for fresh evaluation (squash prep) and formalin fixation.

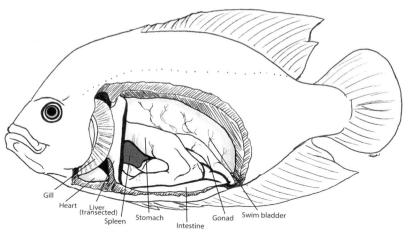

Figure 8.16 Schematic drawing of an oscar (*Astronotus ocellatus*) dissection with some of the major organs labeled after the liver has been transected. Drawing by K Cook

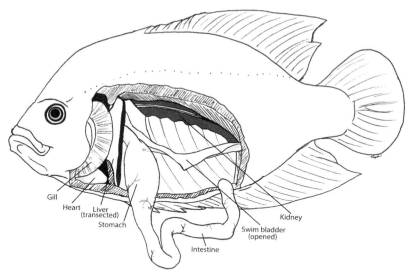

Figure 8.17 Schematic drawing of an oscar (*Astronotus ocellatus*) dissection with most of the major organs labeled after the liver has been transected and the swim bladder opened. Drawing by K Cook

- Wet mounts of internal tissues are very useful and, in some instances, may provide more rapid insights and/or identification of organisms more easily than through histopathology (see **Figures 8.21–8.30**). Only small representative pieces of each tissue – small enough to be "squashed" in a drop of water under a cover slip to allow light penetration – are needed. Lesions should be targeted if observed in the following:

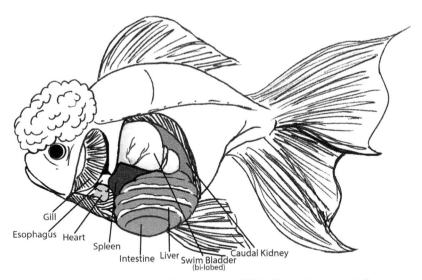

Figure 8.18 Schematic drawing of an oranda goldfish (*Carassius auratus*) dissection with all of the major organs labeled. Drawing by K Cook

- Liver, spleen, and anterior and posterior kidney.
- Stomach and intestinal tract (if tissues are too thick, it will be difficult to assess on wet mount; it may be necessary to evaluate scrapes of stomach and intestine for parasites and then cut thinner sections to evaluate GI tract layers).
- Gonads for evaluation of gamete reproductive stage and breeding status.
- Any other lesions observed in/around the heart or within organs in the body cavity/mesentery.
- If multiple fish are affected with similar clinical signs are available, some fish can be used for culture, wet mounts, virology, or other molecular diagnostics, and others can be used for histopathology.

NORMAL FISH ANATOMY (FIGURES 8.12–8.20)

- The mouth and oral cavity vary among species. Many have jaw teeth of varying shapes and configurations depending on native diets, but a number of common species also have pharyngeal (throat) teeth and/or pads (e.g., moray eels, cichlids, koi) that assist with further break down or capture of food.
- The gill is the primary respiratory organ for fish and is responsible for both oxygen uptake and carbon dioxide and ammonia removal from the blood.

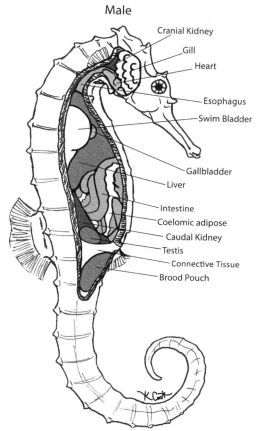

Figure 8.19 Schematic drawing of a male seahorse (*Hippocampus* sp.) dissection with all of the major organs labeled. Drawing by K Cook

The oral and opercular cavity serve as the respiratory pump that directionally drives water from the mouth over the gills and past the operculum to facilitate efficient countercurrent exchange of these gases. Water must move in this direction. The gill is also a site of osmoregulation.
- In many species, a gill-like structure, called the pseudobranch, often near or on the inner surface of the operculum, may be visible inside the gill chamber. The pseudobranch is believed to be important for gas exchange to the eye.
- The heart is two-chambered (one ventricle and one atrium). A thin sinus venosus (often white) leads, via the sinoatrial canal and sinoatrial valve (SAV), into the atrium which is thin and sac-like. The atrioventricular (AV) valve allows blood from the atrium to enter the ventricle, which is typically

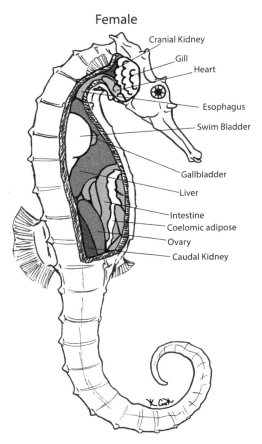

Figure 8.20 Schematic drawing of a female seahorse (*Hippocampus* sp.) dissection with all of the major organs labeled. Drawing by K Cook

thicker and either tubular, pyramidal, or sac-like in shape. Shape, thickness, and size varies among species. A more elastic bulbus and conus arteriosus (current understanding is that both are found, to varying degrees, in many bony fish and elasmobranchs) receive blood from the ventricle and normalize pressure and flow to the rest of the circulatory system.
- The liver has functions similar to that of other vertebrates. Liver size, shape, and color vary depending on diet, age, sex, and reproductive status. Livers are often tan- to salmon- colored. Well-fed fish and/or reproductively active females may have more fat or yolk precursors affecting coloration. Over-fed fish may have very white or pale tan livers, suggesting excess fat/hepatic lipidosis. Bony fishes often have pancreatic tissue interspersed within the liver

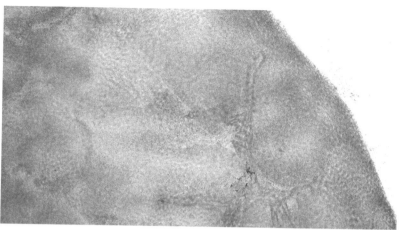

Figure 8.21 Normal liver wet mount from a firemouth cichlid (*Thorichthys meeki*). Normal liver wet mounts may vary depending upon condition of fish, sex, and reproductive stage. RP Yanong

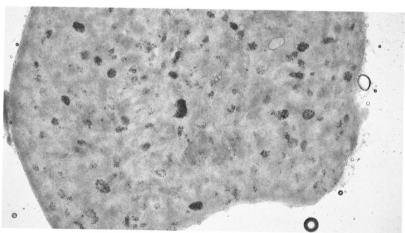

Figure 8.22 Normal spleen wet mount from a firemouth cichlid (*Thorichthys meeki*). Note red and white pulp and pigmented macrophage aggregates. RP Yanong

as well as in mesentery and other areas of the coelom. Some species (including, e.g., elasmobranchs) have a more discrete pancreas.
- The GI tract of fish varies greatly. As discussed earlier, some species (e.g., cyprinids including koi and goldfish, pipefishes, parrotfishes) lack a true, acid-producing, saccular stomach. Stomachs (in species with one) vary in size, shape, and muscularity. Some have pyloric pouches or cecae (e.g., neon tetra), which increase absorption and digestive capacity, located at

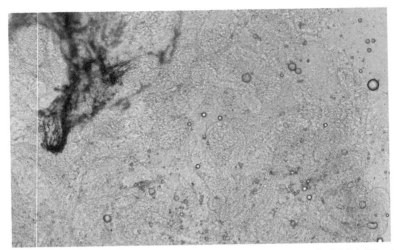

Figure 8.23 Normal posterior kidney wet mount from a firemouth cichlid (*Thorichthys meeki*). Tubules and glomeruli can be seen. RP Yanong

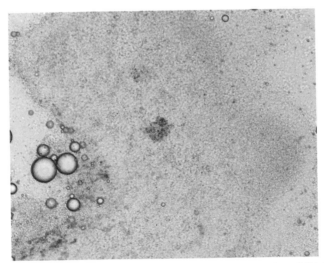

Figure 8.24 Normal anterior kidney wet mount from a firemouth cichlid (*Thorichthys meeki*). Note pigmented macrophage aggregates and general homogenous brownish appearance. RP Yanong

the junction of their stomach and intestinal tracts. The length and internal structure of the intestine varies as well depending on fishes' natural diet. In general, more herbivorous fishes have longer GI tracts, and more piscivorous/carnivorous fishes have shorter tracts. Omnivore GI lengths are somewhat intermediate.

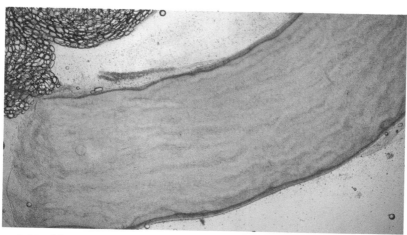

Figure 8.25 Normal intestine wet mount from a firemouth cichlid (*Thorichthys meeki*). Note intestinal folds. RP Yanong

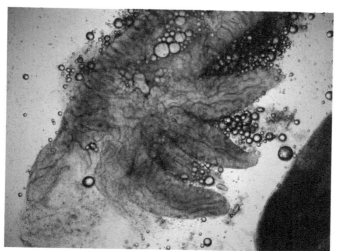

Figure 8.26 Pyloric cecae from a neon tetra (*Paracheirodon innesi*). M Hyatt/J Meegan/ RP Yanong

- Kidneys vary in shape depending on the species but, in general, are dark red in coloration and divided into an anterior kidney – which functions primarily as a hematopoietic, immune, and endocrine organ, and posterior kidney – which functions primarily as an excretory organ. Grossly, the kidney can be a long single, fused organ (trout), have more excretory kidney function concentrated anteriorly (some poeciliids), or can be divided into two regions (anterior and posterior in catfish, koi, goldfish). The left and right kidneys

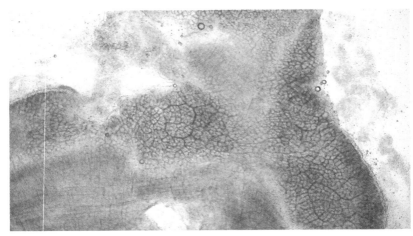

Figure 8.27 Normal stomach wet mount from a firemouth cichlid (*Thorichthys meeki*). Note "honeycomb-like" gastric glands. RP Yanong

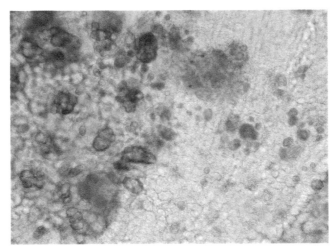

Figure 8.28A Stomach with numerous granulomas from cichlid with *Cryptobia iubilans* infestation. Note presence of "honeycomb-like" gastric glands identifying stomach tissue. RP Yanong

are fused to a varying extent (species dependent) and lie retrocoelomically, dorsal to the swim bladder and ventral to the spine in most species.
- The spleen is varied in size and shape in fish, but is commonly a red round, triangular, to elongate organ often near the stomach. It has roles in hematopoiesis, immunity, and red blood cell destruction/recycling.
- Swim (air or gas) bladders:

Normal Fish Anatomy (Figures 8.12–8.20)

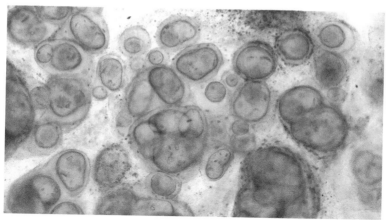

Figure 8.28B Close up of granulomas from cichlid with *Cryptobia iubilans* infestation. RP Yanong

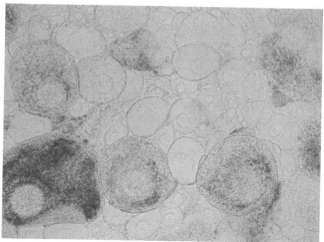

Figure 8.29 Normal ovarian tissue wet mount from a gold variety gourami (*Trichopodus trichopterus*). Note oocytes of different stages of maturation present including clearer primary oocytes. M Hyatt/J Meegan/ RP Yanong

- Most bony fish (most pet fish, except sharks and rays) have a swim bladder (i.e., gas bladder, air bladder) that regulates buoyancy. Some species that tend to be more bottom dwelling (benthic/demersal), such as marine blennies, gobies, and hawkfish, lack a swim bladder. Marine sharks and rays use their lipid-rich liver to assist with buoyancy.
- Fish species that, as adults, have swim bladders connected to the esophagus via a pneumatic duct (including koi and goldfish) are considered

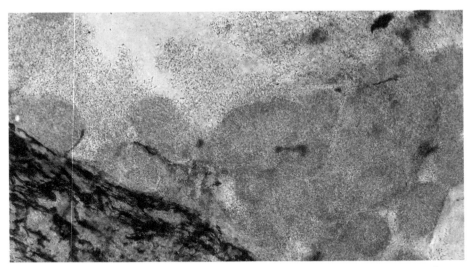

Figure 8.30 Biopsy wet mount from normal testis of a live bearing poeciliid reveals a cluster of round to ovoid spermatozeugmata (sperm packets that are transferred into the livebearer female via the male gonopodium during breeding), including some that have been burst open to show sperm. RP Yanong

physostomous. Those without this connection as adults (such as cichlids) are physoclistous. Although this connection can help with adult swim bladder inflation, physostomous fish may also have swim bladder issues related to this esophageal connection.
- Swim bladders can be a single chamber (e.g., cichlids) or two chambers connected by an isthmus (carp, goldfish, tetras) or more, and even more complex. Swim bladders may be opaque and thick or clear and thin. In many, a vascular rete and a gas gland may be apparent at necropsy. Both of these help control gases in the swim bladder and assist with buoyancy control.
- The sensory lateral line system can be observed as a line of dots or pits along and around the head and lateral body scales of the fish and mediates responses through the central nervous system (CNS). In some evolutionarily older "bony fishes" (e.g., gar, sturgeon, paddlefish) as well as in elasmobranchs, electro-sensory pores, known as ampullae of Lorenzini, are present around the head/mouth region.

CHAPTER 9
EUTHANASIA, ANESTHESIA, AND SEDATION

GENERAL CONSIDERATIONS FOR EUTHANASIA

- Use tank/system water (if parameters are within acceptable ranges).
- Reduce environmental stimuli (noise, use of opaque container).
- Have adequate aeration (air stone, pump).
- Mix the drug with some of the system water prior to adding it to the container with fish, or move the fish into a container with the euthanasia drug already added.
- *For more information, see the most current edition of the *AVMA Guidelines for the Euthanasia of Animals* (which is the primary reference for information in this chapter).

INDICATORS OF DEATH IN FISH

- Loss of movement, loss of reactivity to any stimulus, and initial flaccidity (prior to rigor mortis).
- Respiratory arrest (cessation of rhythmic opercular activity) for a minimum of 30 minutes and loss of eyeroll (vestibulo-ocular reflex, the movement of the eye when the fish is rocked from side to side).
- Presence of a heartbeat is not a reliable indicator of life, but the absence of a sustained heartbeat suggests death has occurred.
- Secondary methods of euthanasia may be necessary for added certainty.
- See **Table 9.1** for stages of anesthesia in fishes.

EUTHANASIA METHODS

For most pet fish owners, immersion or injectable will be the most acceptable methods (i.e., vs physical methods). The immersion methods described in the next section can be used as a one-step procedure but not until after a minimum of 30 minutes after cessation of opercular movement/respiration; however, in some instances for any of the one-step procedures, a secondary step (e.g., pithing or intracardiac pentobarbital) may be warranted, especially for hypoxia tolerant species like goldfish, catfish, and cichlids.

Immersion Methods
- Tricaine methanesulfonate (MS-222, buffered) (one step): 250–500 mg/L, or up to 5–10 times the anesthetic dose for 30 minutes after cessation of opercular movement/respiration. Should be buffered with sodium bicarbonate (1:1 by weight) or sodium percarbonate (MS-222:buffer, 0.5:1 by weight).
  - Koi, goldfish, and cichlids require a secondary step, i.e., pithing or intracardiac pentobarbital, if MS-222 is used.
- Metomidate hydrochloride (one step): 100–250 mg/L, for 30 minutes after cessation of reactivity to vibration or touch (~ 45–70 minutes total); a few species (e.g., *Corydoras* catfish) may require up to 1000 mg/L. Fish demonstrate reactivity to vibration and touch even after cessation of respiration. Should be buffered to system water pH with sodium bicarbonate or sodium carbonate.
- Eugenol, isoeugenol, and clove oil (one step): anesthetic concentrations will vary depending on species and other factors, but it may be as low as 17 mg/L for some species. Greater concentrations (10 times the upper range for anesthesia) will be required for euthanasia. Oils are not very water soluble; injecting the solution through a syringe and fine-gauge needle under the water in the euthanasia container will help ensure it is dispersed in the water. Fish should be left in the anesthetic solution for a minimum of 30 minutes after cessation of opercular movement.
- Ethanol (one step): 10–30 mL of 95% ethanol/L (= 12.7 to 38 mL of 75% ethanol) prolonged immersion (30 minutes after cessation of opercular movement).

Injectable Methods
- Pentobarbital (one step): sodium pentobarbital (60–100 mg/kg [27.3 to 45.5 mg/lb]) can be administered intravenously (IV) or by intracardiac or intracoelomic routes. Pentobarbital may also be administered via intracardiac injection for anesthetized animals as the second step of a two-step euthanasia procedure. Death usually occurs within 30 minutes.
- Ketamine-medetomidine (two step): a combination of ketamine, at dosages of 1–2 mg/kg, with medetomidine, at dosages of 0.05–0.1 mg/kg (0.02–0.05 mg/lb), may be administered via intramuscular (IM) injection followed by a lethal dose of pentobarbital.

- Propofol (two step): a dose of 1.5–2.5 mg/kg (0.7–1.1 mg/lb) can be administered by IV, followed by an injection of a lethal dose of pentobarbital (see above).
- Potassium chloride (KCl) can be used at 10 meq/kg IV or intracardiac as a secondary method to stop the heart.

ANESTHESIA AND SEDATION

Tricaine Methanesulfonate (Syncaine®, MS-222)
- Make a stock solution with 10 g/L of dechlorinated, deionized water (using a special filter) or distilled water.
- Store away from light (in an amber bottle) at 4°C; the mixture is potent for up to 6 months (Katz et al., 2020).
- Must buffer with sodium bicarbonate (approximately 1:2 or 1:1, MS-222 to sodium bicarbonate in grams) or sodium carbonate (0.5:1 or 1:1) prior to use (see Chapter 14, **Table 14.3**).
- Concentrations of 100–150 mg/L will sedate or anesthetize many species of fish in 3–5 minutes (100 for most fish, 150 for larger cichlids, koi). Some species, e.g., gouramis, may take higher concentrations (up to 300–400 mg/L).
- Higher doses take effect more quickly and may reduce the excitatory phase; lower doses are appropriate for maintenance or long term.

Clove Oil (A Mixture of Isoeugenol, Methyleugenol, and Eugenol)
- Available at pharmacies.
- Dilute in 95% ethanol at 1:9 ratio to make 100 mg/ml (assuming that each mL of pure eugenol contains 1 g of active drug).
- Concentration of 25–120 mg/L will anesthetize fish; the recovery is more prolonged than with MS-222. Ten times the upper anesthetic dose range will be required for euthanasia.

ANESTHESIA TIPS
- Anesthesia stages: excitatory, sedation, loss of equilibrium, loss of any reactivity, and death. See **Table 9.1**.
- Anesthesia is adequate if the fish has only mild/slow opercular movements, relative to unsedated "normal" opercular rate (observe and quantify rate prior to anesthesia).

Table 9.1 **Stages of Anesthesia in Fishes**

STAGE	PLANE	DESCRIPTION	SIGNS
0	0	Normal	Swimming actively, equilibrium normal
I	1	Light sedation	Reduced motion, ventilation decreased
I	2	Deeper sedation	Motionless unless stimulated
II	1	Light anesthesia	Partial loss of equilibrium
II	2	Deep anesthesia	Total loss of equilibrium
III	1	Surgical anesthesia	Total loss of reactivity when stimulated
III	2	Deep surgical anesthesia	Decrease in respiratory and heart rates
IV	1	Medullary collapse	Cessation of respiratory movements
IV	2	Cardiac arrest	Death

Saint-Erne (2019), modified from Brown (1993) and Ross (2001).

- Obtain longer term anesthesia using a recirculating anesthesia machine, ensuring gills are continuously bathed with anesthetic water (90–160 mg/L, 3 L/minute).
- Reverse or reduce anesthetic level by reducing anesthetic concentration (adding water with no drug) or placing fish in fresh clean water.
- After anesthesia and procedures are complete, place fish in aerated recovery tank.
- If manual/assisted recovery is deemed necessary, make sure water flows from the mouth out the opercula (do not recover the fish using a back-and-forth movement; only forward movement or with water circulating in through the mouth).

CHAPTER 10
SURGERY

COMMON INDICATIONS FOR SURGERY

- Correction of swim bladder abnormalities.
- Removal of growths (neoplastic, inflammatory, or parasitic).
- Wen trimming. The wen is a gelatinous growth or crown on the head and face of many fancy goldfish varieties.
- Gonadectomy (e.g., gonadal disease, including neoplasia).
- Foreign body removal.
- GI impaction relief.
- Enucleation.
- Endoscopic/exploratory exam for sexing or diagnosis.
- Implanting telemetry devices.
- See **Table 10.1** (Surgical Supplies and Equipment).

FISH ANESTHESIA DELIVERY SYSTEM (FADS)

Although some surgical procedures can be fairly rapid, longer procedures may require the use of a fish anesthesia delivery system (FADS). There are many different designs for a FADS, both simple and more complex. A relatively simple system is described in this section (see **Figures 10.1–10.3**).

An effective and economical recirculating fish anesthesia system requires the following supplies.

- Commercially available submersible power head.
- Plastic tubing and clamps from a hardware store.
- For medium to larger fish, an airline tube "Y" connector can be used (with additional airline at each "arm") to help direct water flow to each set (left and right) of gills.
- 10-gallon aquarium.
- Custom-made acrylic support or modified tank cover.
- Open-cell foam V-tray cut to fit the patient (can wrap in plastic).

DOI: 10.1201/9781003057727-10

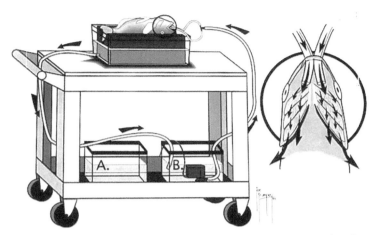

Figure 10.1 An early fish anesthesia delivery system (FADS) using tanks of water and a submersible pump. Lewbart et al., 1995

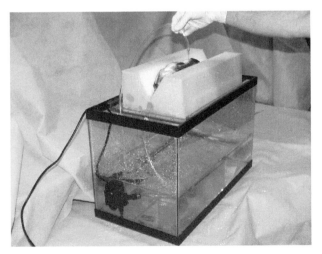

Figure 10.2 A self-contained fish anesthesia delivery system (FADS) utilizing a 10-gallon aquarium. A plastic sweater box or similar can be used in place of the bulky glass tank. Lewbart and Harms, 1999

- Primary flow of the anesthesia-laden water is delivered through the mouth to the gills (unidirectional), and trickles down through the foam to the aquarium reservoir for recirculation.
- A secondary flow can be diverted to keep the skin moist.

FISH PREPARATION

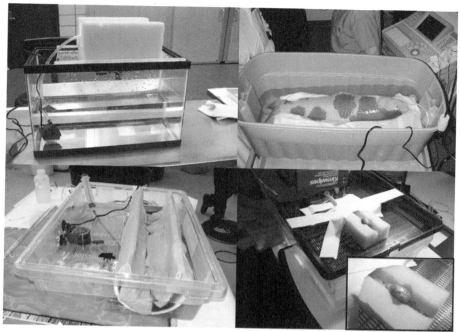

Figure 10.3 A variety of FADS setups. Clockwise from upper left (simple FADS utilizing a glass aquarium, plexiglass platform, closed foam support, and a submersible pump; large koi, *Cyprinus rubrofuscus*, on a FADS made from a large Rubbermaid® container; a 1 gram zebra danio, *Danio rerio*, anesthetized for an abdominal exploratory utilizing a critter cage and a fluid intravenous set to deliver the anesthetic; a plastic tub with a PVC trough and chamois cloth for anesthetizing small elasmobranchs). GA Lewbart

INDUCTION, MAINTENANCE, MONITORING

- Fish anesthetic stage/plane is best monitored based on respiration (rate of opercular movement, i.e., rate of gill plate or operculum) and refer to **Table 9.1**.
- Use of a Doppler to monitor heart rate may be helpful (see **Figure 10.4**).
- MS-222 is the most commonly used anesthetic for surgical procedures on fish.
- Other agents that are regularly used and have merit are eugenol, propofol, alfaxalone, 2-phenoxyethanol, and metomidate among others.

FISH PREPARATION

- Pre-operation: evaluate with radiology and/or ultrasonography.

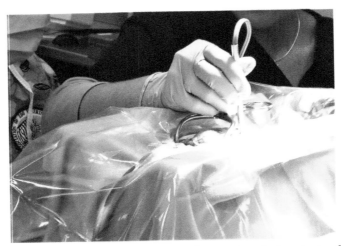

Figure 10.4 A Doppler pencil probe being used to monitor the heart rate of a surgical patient under anesthesia. RP Yanong

Table 10.1	Surgical Supplies and Equipment
Supplies: • Surgery pack (with appropriate instruments) • Suture absorbable and non-absorbable (monofilament) • Cotton-tipped swabs • Clear plastic drape • Sterile saline • Surgery sponges (2x2 or 4x4) • Scalpel blades • Surgical gloves (selection of sizes) • Masks • Caps • Scrub tops • Head loupes • Gelpi retractors • +/−Doppler flow probe • +/− ECG • Medications: • Antibiotics (injectable) • Butorphanol (at 0.1–0.4 mg/kg IM/SQ) • Ketoprofen • Triple antibiotic ointment • Povidone iodine ointment • Silver sulfadiazine cream • Sterile ophthalmic lubricant • Salt (1–3 ppt in recovery tank) for freshwater fishes	

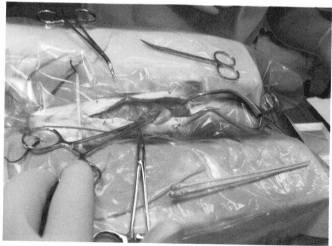

Figure 10.5 Plastic surgical drapes have the advantage of being see-through for patient monitoring and keeping moisture in. RP Yanong

- Withhold food for 12–24 hours prior to surgery.
- Anesthetize the fish, then place on a wet foam pad with a "V" cut out to support the fish.
- Keep the fish moist with regular basting.
- Remove scales from the surgical area with forceps.
- Wipe incision area with a small amount of sterile saline, dilute povidone iodine (1:20), or chlorhexidine (1:40).
- Routine pre-operative antibiotics may reduce secondary bacterial infections (enrofloxacin 10 mg/kg ICe or ceftazidime 20 mg/kg IM).
- Clear plastic drapes retain moisture and provide a sterile field (see **Figure 10.5**).

SURGICAL TIPS

- Bipolar cautery helps to control hemorrhage (use caution in small patients to avoid damaging adjacent tissues). Bipolar cautery is not a problem with a marine anesthetic system.
- Use small animal instruments and/or ocular or microsurgical pack and head loupe magnification.
- When removing growths, strive for wide margins over complete primary closure.
- Swim bladders are fragile and can collapse if torn.

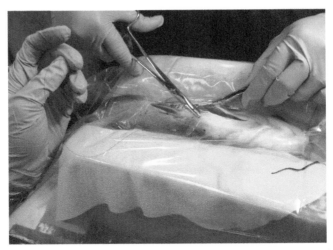

Figure 10.6 Closing the body wall of a surgical patient. Here, an assistant helps manage the suture material. RP Yanong

- "Absorbable" sutures may not be absorbed in fish; use monofilament to avoid bacterial colonization. Polydioxanone sutures have the shortest healing time and cause the least amount of skin reaction. Cyanoacrylate irritates and may inhibit wound/incision closure and, therefore, should be avoided.
- Use a cutting tip needle for sutures applying one of the following patterns: simple continuous, simple interrupted, or Ford interlocking pattern on skin, with single or two-layer closure (see **Figure 10.6**).
- The integument is generally the holding layer. Aspirate air from coelom with suction when closing the muscular layer.
- Control pain with butorphanol, ketoprofen, or another appropriate analgesic agent.
- Lack of mobile skin makes defects difficult to close; second-intention healing is common. Reduce osmotic gradient to enhance healing: add salt to pond/tank at 1–3 g/L for freshwater patients.
- Apply povidone iodine ointment to closed incision before returning fish to the recovery water.
- Remove skin sutures upon healing (generally 10–14 days but time to suture removal may be longer in some species and at lower water temperatures).

CHAPTER 11
INFECTIOUS DISEASES

GENERAL MANAGEMENT CONSIDERATIONS

Multiple non-infectious and infectious disease factors often contribute to a disease outbreak. A complete history combined with evaluation of the system and water quality are as important as the physical exam and subsequent diagnostics. Environmental or management problems often facilitate infectious diseases, so these must be remedied prior to or concurrent with any treatments.

The following are general considerations when using any drugs or chemicals:

- Microbiome and nitrifying bacteria. More and more research shows the importance of microbiota both in the environment and in and on the fish for maintaining health. Any use of chemicals or drugs, especially when used as an immersion treatment, may change the microbiome and nitrifying bacteria – as well as lead to drug resistance – so it is important that these are used judiciously.
- Hospital tank. Isolating sick fish into a hospital tank with a secure lid for biosecurity is preferred to reduce disease spread and minimize drug and chemical effects on the origin system. Provide shelter or other structures to reduce stress, and have dedicated nets and other supplies for each tank. Ideally, immersion antibiotics should be provided in a separate hospital tank that can be managed and cleaned easily. For some diseases and treatments (e.g., immersion), if adequate water is available, treatments followed by 70%–100% water changes should suffice. For other diseases and management (e.g., injectable, oral), an easily cleaned filter (e.g., sponge filter) will help manage ammonia.
- Hospital tank disinfection. Hospital tanks should be cleaned and disinfected as well as the filtration changed out between fish cases. A separate tank can be used to "cycle" other sponge filters with nitrifying bacteria (and fed ammonia) so that a biofilter will always be available for new cases unless the tank will be managed with water changes alone.

DOI: 10.1201/9781003057727-11

PREVENTIVE CARE: VACCINES

- There are two major types of vaccines:
- Commercially available.
- Autogenous vaccines.
  - Commercially available fish vaccines are designed primarily for food fish species. Few to none are relevant for common aquarium display fish.
    - Available commercial vaccines: https://www.aphis.usda.gov/animal_health/vet_biologics/publications/aquaproducts.pdf
  - Autogenous vaccines are made based on clinical isolates from and for a specific facility. Currently, autogenous vaccine manufacturers are more geared toward larger volumes needed for commercial production facilities than for smaller scale needs, although this may change in the near future.

PARASITIC DISEASES OF AQUARIUM FISH

Parasites are a very common contributor to fish disease outbreaks (see **Tables 11.1–11.6** and **Figures 11.1–11.34**). Microscopic wet mount evaluation of skin, fin, and gills are an important diagnostic standard.

Parasites can be divided into one-celled protistan parasites and multi-cellular (and typically larger) metazoan parasites. Protistan parasites include ciliates (the most common protistan parasites of fish), flagellates, dinoflagellates, microsporidia, and coccidia.

Metazoan parasites include "insect-like" and "worm-like" parasites. "Insect-like," i.e., crustacean parasites, have somewhat analogous life stages to arthropod parasites of domestic (terrestrial) species and are managed with drugs similar to those used against them. Crustacean parasites include parasitic copepods, branchiurans, and isopods. "Worm-like" parasites share a vermiform appearance but are not all related. In fact, some are not "true" worms. The true "flatworms" (Platyhelminthes) include monogeneans (the most common metazoan parasites of fish), digenean trematodes, cestodes, and turbellarians. Another common group are the nematodes (roundworms). Less common, but seen occasionally, are leeches and pentastomes (nymph life stages seen in the coelom of fish).

Understanding the life cycles of these parasites, as for those seen in terrestrial species, are critical for proper management.

Table 11.1 Parasites: Protists - Motile Ciliates

NAME, (FRESH (F), SALT (S), BRACKISH (BR)), APPROXIMATE SIZE	TARGET TISSUE(S)	CLINICAL SIGNS	MOVEMENT/ FEATURES	LIFE CYCLE: DIRECT (D)/ INDIRECT (I), ADDITIONAL INFO	PREDISPOSING FACTORS	MANAGEMENT	NOTES
Ichthyophthirius multifiliis (F) (FW white spot) theront: 30–50um; trophont: 50um–1mm	Skin, fin, gills	White spots, increased mucus, tachypnea	Trophont feeding stage) – slow, rolling/amoeboid- round, various sizes, gray granular, C-shaped macronucleus (not always visible); theront pyriform	D – stages on and off the host	Rapid temperature drop/immune suppression	Formalin, Salt (4–5 ppt)	Life cycle duration and Tx temperature dependent; 25°C (77°F): treat 5–10 days; stages on/off host
Cryptocaryon irritans (S) (SW white spot) theront: 25–60um	Skin, fin, gills	White spots, increased mucus, tachypnea	Somewhat similar to *Ichthyophthirius* but more opaque cytoplasm (nucleus difficult to see)	D – stages on and off the host	Temps >19°C (66°F)	Copper, chloroquine	Life cycle duration and Tx temperature dependent; 25°C (77°F): treat 4–6 weeks; stages on/off host
Trichodina/ Trichodonella, Tripartiella (F, S, Br) (16–111um)	Skin, fin, gills, some internal	Increased mucus (gray/ blue hue)	Radial symmetry, "scrubbing bubble" or "flying saucer"	D	High organics	Reduce organic loading, Formalin, potassium permanganate	–

(Continued)

Table 11.1 (Continued)

NAME, (FRESH (F), SALT (S), BRACKISH (BR)), APPROXIMATE SIZE	TARGET TISSUE(S)	CLINICAL SIGNS	MOVEMENT/ FEATURES	LIFE CYCLE: DIRECT (D)/ INDIRECT (I), ADDITIONAL INFO	PREDISPOSING FACTORS	MANAGEMENT	NOTES
Tetrahymena (F) (30 x 55um)	Skin, fin, gills, muscle – can go systemic	Erosion, ulceration	Football/pear shaped, with large granules	D – but can form 2–4 tomites per adult	High organics, damage to epithelium	Formalin, formalin/ malachite green	May infest fish post-mortem; live/moribund fish better for diagnostics
Uronema (S, Br) (30–50um)	Skin, fin, gills, muscle – can go systemic	Erosion, ulceration	Football/elongate, movement similar to *Tetrahymena*	D	High organics, damage to epithelium	Formalin	May infest fish post-mortem; live/moribund fish better for diagnostics
Chilodonella (F) (30–80um x 20–62um)	Skin, fin, gills	Increased mucus, tachypnea, emaciation	Smooth, gliding movement; ovoid, kidney shaped, flat, bands of cilia, clear	D	Environmental/ handling stressors	Formalin, salt (3–5 ppt)	–
Brooklynella (S) (36–86um x 32–50um)	Gills primarily	Respiratory distress	Flat kidney/heart shaped, clear	D	Shipping, environmental stressors	Formalin	–

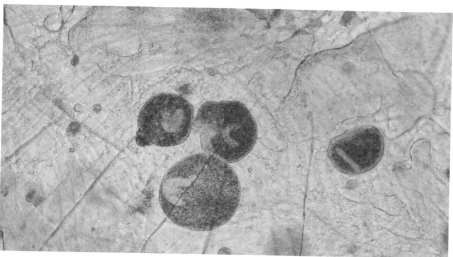

Figure 11.1 A large number of *Ichthyophthirius multifiliis* (Ich) organisms are seen in this skin scrape. The pale, crescent-shaped macronucleus can be observed in some of the mature trophozoites. RP Yanong

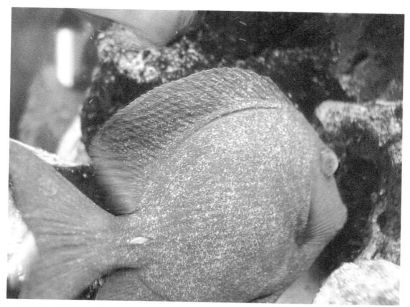

Figure 11.2A This Atlantic blue tang (*Acanthurus coeruleus*) has numerous small white spots commonly seen with *Cryptocaryon irritans* infestation. D Van Genechten

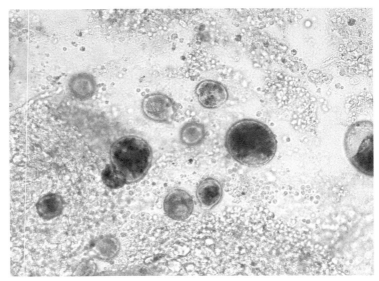

Figure 11.2B   A large number of *Cryptocaryon irritans* organisms are seen in this skin scrape of a mandarin goby (*Synchirops splendidus*). Unlike ich, the nucleus is not very prominent in most cases. RP Yanong

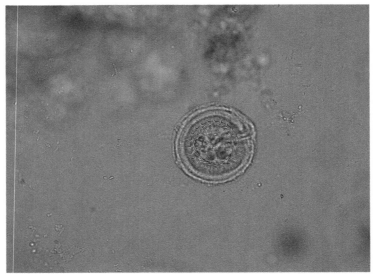

Figure 11.3   One trichodinid, a ciliated protozoal parasite with radial symmetry, is seen in this micrograph. G Trende

Parasitic Diseases Of Aquarium Fish

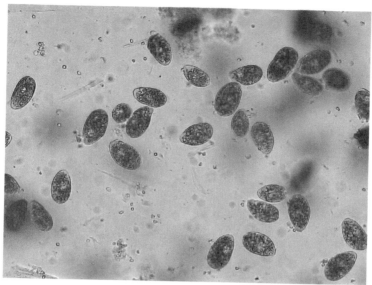

Figure 11.4 A group of *Tetrahymena* sp., a three-dimensionally ovoid ("egg-shaped"), ciliated protozoal parasite with bilateral symmetry, is seen in this micrograph. G Trende

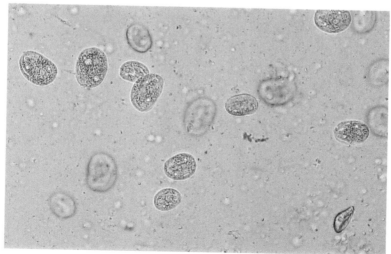

Figure 11.5A A group of *Chilodonella* sp., a flattened and ovoid ("potato-chip shaped"), freshwater ciliated protozoal parasite, is seen in this micrograph. G Trende

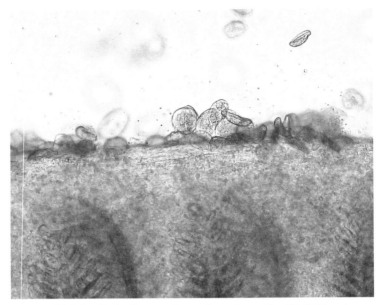

Figure 11.5B *Chilodonella* feeding on the surface of a molly gill. G Trende

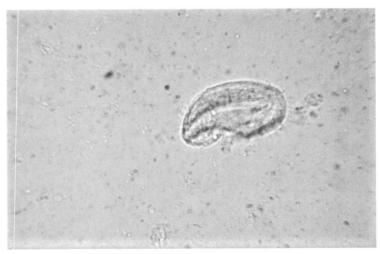

Figure 11.6 One *Brooklynella* sp., an ovoid to reniform marine ciliated protozoal parasite, is seen in this micrograph. D Pouder

Table 11.2 Parasites: Protists - Sessile Ciliates

NAME	TARGET TISSUE(S)	CLINICAL SIGNS	MOVEMENT/ FEATURES	LIFE CYCLE (BINARY FISSION)	PREDISPOSING FACTORS	MANAGEMENT	NOTES
Epistylis (F) (40–80um x 20–30um)	Skin, fin, gills	If large colony size, small tufts or nodules on fin, skin, gills;	Colonial; elongate bell-shaped zooids with ciliated oral end, on stalk which is up to 1.2mm length	D	High organics, benthic lifestyle	Reduce organic loading; formalin, potassium permanganate, 4–5 g/L salt	Can be associated with motile aeromonad infection ("red sore disease")
Ambiphrya/ Scyphidia (Br, F) (44–95um x 24–61um)	Skin, fin, gills	—	Barrel-shaped, oral and midline rows of cilia	D	High organics	Reduce organic loading; formalin, potassium permanganate, 4–5 g/L salt	—
Apiosoma (F) (40–110um x 22–40um)	Skin, fin, gills	—	Solitary; elongate and tapered to cup-shaped	D	High organics	Reduce organic loading; formalin, potassium permanganate, 4–5 g/L salt	—
Capriniana (F) (31–99um x 21–75um)	Gills, skin	Tachypnea	Amorphous shape, has pincushion-like appearance (suctorial tentacles)	D – by budding	High organics	Reduce organic loading; formalin, potassium permanganate, 4–5 g/L salt	Feeds on other ciliophorans, so need to reduce this food source

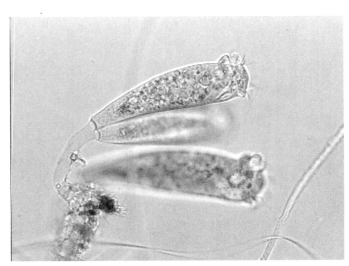

Figure 11.7   Several *Epistylis* sp., a stalked, ciliated protozoal parasite, are seen in this micrograph. RP Yanong

Figure 11.8   One *Ambiphyra* sp., a stalked, barrel-shaped ciliated protozoal parasite, is seen in this micrograph. D Pouder

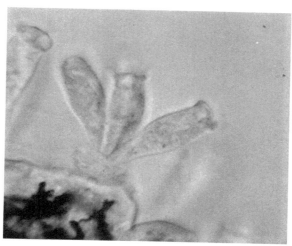

Figure 11.9 Several *Apiosoma* sp., a stalked, ciliated protozoal parasite, are seen in this micrograph. RP Yanong

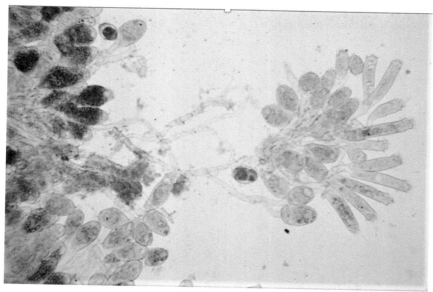

Figure 11.10 A large number of *Heteropolaria* sp., a stalked, ciliated protozoal parasite, are seen in this micrograph. GA Lewbart

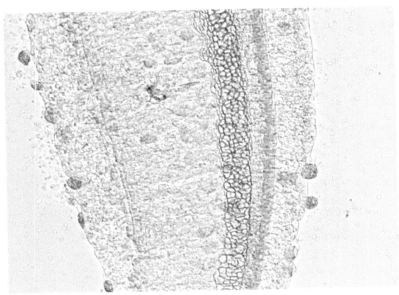

Figure 11.11A Several *Capriniana* sp., a stalked, ciliated protozoal parasite, are seen on the edges of a gill in this micrograph. L Khoo

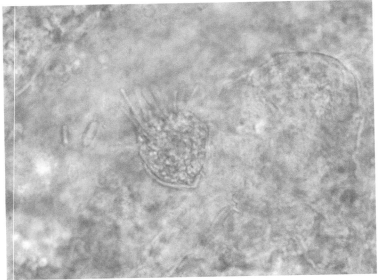

Figure 11.11B Higher magnification micrograph of *Capriniana* sp. on a gill. L Khoo

Table 11.3 Parasites: Protists - Flagellates

NAME	TARGET TISSUE(S)	CLINICAL SIGNS	MOVEMENT/ FEATURES	LIFE CYCLE INFO	PREDISPOSING FACTORS	MANAGEMENT	NOTES
Ichthyobodo spp. (F, S, Br, sp. dependent) (5–20um x 3–15 um)	Skin, fin, gills	Increased mucus, flashing, tachypnea	Can be attached and unattached; ovoid/ teardrop to round shaped; "flickers" when attached	D – asexual binary fission	High density, poor nutrition, higher temp	–	May see dividing stage with 3–4 flagella
Cryptobia branchialis (F, S, Br) (14–23um x 2.2–4.8um)	Gill	Tachypnea	"Wiggles"	D – asexual binary fission	–	Formalin	–
Spironucleus sp. (F, S, Br) (12.5–20.5um x 5.0–11.2um)	Intestine; can spread to other tissues	Anorexia, weight loss, coelomic distension, darkening, lethargy, abnormal swimming, buoyancy, hanging in the corner, less common – external lesions	Pyriform-shaped, directed, rapid movement	D – asexual, binary fission; can reproduce in 2 hours	Handling or environmental stressors; in feces; cannibalism	Reduce stressors, metronidazole – oral or immersion	Very common in cichlids; this or similar species can be seen in other tropical fish species

(*Continued*)

Table 11.3 (Continued)

NAME	TARGET TISSUE(S)	CLINICAL SIGNS	MOVEMENT/ FEATURES	LIFE CYCLE INFO	PREDISPOSING FACTORS	MANAGEMENT	NOTES
Cryptobia iubilans (F) (5.5–12.5um x 3.5–5.5um)	Stomach; can spread to other tissues	Inappetence, lethargy, hanging, darkening, tachypnea	"Wiggles"	D – asexual binary fission; two phases: 1) elongate phase (pankinetoplastic); 2) stout, tear-drop shaped (eukinetoplastic); can migrate to other sites	Environmental stressors	80 mg/L dimetridazole, 24 hours x 5–7 days	Feces, cannibalism, can survive in water column for a few hours; live parasites not always seen on wet mount; granulomatous gastritis common; can go systemic in extreme cases
Protoopalina spp. (F, S) 62–278 um x 15–42 um	Intestine	No signs, weakened, lethargic	Relatively large, elongate, tapered to a point on one end, rounded on the other; numerous "flagella" (which look like "cilia")	D	–	Metronidazole appears to help	Considered endosymbionts in some species, e.g., some marine angelfish species

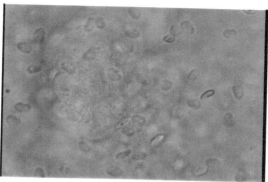

Figure 11.12 A large number of *Ichthybodo* sp., a small, flagellated parasite, are seen in this micrograph. D Pouder

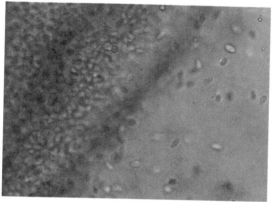

Figure 11.13A A large number of *Spironucleus vortens.*, a small, flagellated parasite, are seen in this micrograph. RP Yanong

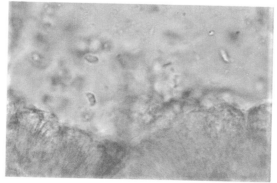

Figure 11.13B High magnification micrograph of several *Cryptobia iubilans* (stouter/teardrop shape/phase) located along the gut mucosa in a cichlid. M Hyatt/J Meegan/RP Yanong

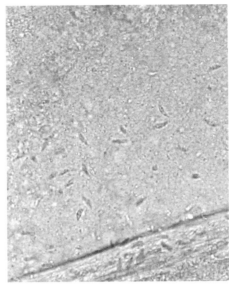

Figure 11.13C Several *Cryptobia iubilans*, which can have two different shapes/phases. The more elongate phase is seen in this high magnification micrograph. RP Yanong

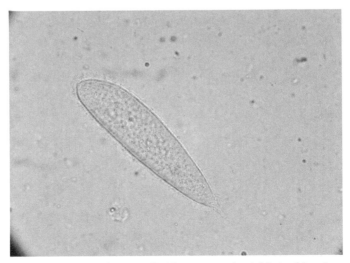

Figure 11.14 A single ciliated protopalinid protozoan is visible in this micrograph. D Pouder

Table 11.4 Parasites: Other Protists

CATEGORY	NAME	FRESHWATER (F), SALTWATER (S), BRACKISH (BR)	TARGET TISSUE(S)	CLINICAL SIGNS/ PATHOLOGY	MOVEMENT/ FEATURES	LIFE CYCLE INFO	MANAGEMENT
Dinoflagellates	*Piscinoodinium* (up to 140um length)	F	Skin, fin, gills	Tachypnea, increased mucus, "dusty" velvet appearance	Trophont (feeding stage on fish) non-motile; yellow green to brown; teardrop to ovoid shape	D – similar to *Ichthyophthirius* with stages on/off host; completed in 10–14 days	Copper, acriflavine, chloroquine
	Amyloodinium (up to 350um length)	S, Br	Skin, fin, gills	Tachypnea, increase mucus, dusty velvet appearance	Trophont (feeding stage on fish) non-motile; pear to ovoid; golden brown	D – similar to *Ichthyophthirius*; up to 256 motile, flagellated dinospores (infective stage) in one tomont	Can tolerate 3–50 ppt/psu; copper, chloroquine
Microsporidia (2–10 um length)	*Pleistophora hyphessobryconis*	Numerous F	Skeletal muscle	Loss of muscle, inflammation, necrosis may appear as asymmetric wasting or cloudy muscle	Non-motile; spores very small; all share "wooden shoe"/"paper clip" shape	D – intracellular	Experimental: toltrazuril, fumagillin, benzamidazoles; classified as relative of true fungi
	Glugea	Mixed	Intracoelomic	Forms xenomas (pseudotumors)		D	
	Fusaspora stethaprioni	F (tetras)	Liver, intestine, other tissues	–		D – disseminated; does not form large cysts	

(Continued)

Table 11.4 (Continued)

CATEGORY	NAME	FRESHWATER (F), SALTWATER (S), BRACKISH (BR)	TARGET TISSUE(S)	CLINICAL SIGNS/ PATHOLOGY	MOVEMENT/ FEATURES	LIFE CYCLE INFO	MANAGEMENT
Coccidia	*Eimeria, Isospora*, others	Mixed	GI and many other tissues	May be incidental; dependent on affected tissues; GI – thin, wasting, low mortalities	Non-motile	D	Coccidiostats somewhat experimental in fish; sulfadimethoxine, other sulfa drugs, amprolium; for intestinal issues, improve husbandry and nutrition
Mesomycetozoea	*Dermocystidium* (3–13um)	Mixed	Skin, fin, eye, other tissues	In tetras, a white worm-like cystic structure containing spores	Non-motile; wet mount squashes of worm-like structure releases spores	D	No treatment, may be self-limiting in some cases
	Ichthyophonus hoferi (variable sizes)	Mixed	Skin, muscle, heart, liver, kidney	"Sandpaper" skin (from spores in skin); granulomatous response in affected tissues	Non-motile; thick walled, spherical resting stage spores surrounded by granulomatous host response	D – spores germinate and form "hyphae" after several hours post-mortem (and on wet mounts)	No treatment

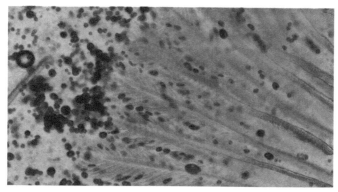

Figure 11.15 A large number of the dinoflagellate *Piscinoodinium* sp. are observed on the gill tissue of this fish. RP Yanong

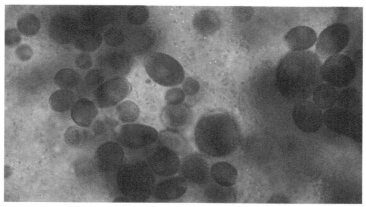

Figure 11.16 A higher magnification of the freshwater dinoflagellate *Piscinoodinium* sp. from gill tissue. RP Yanong

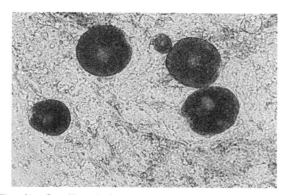

Figure 11.17 Five dinoflagellates belonging to the marine genus *Amyloodinium* are observed in this micrograph. G Vermeer

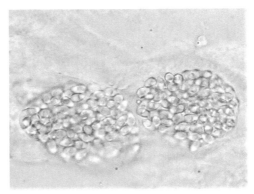

Figure 11.18A This micrograph shows two sporocysts containing numerous spores of the microsporidian *Pleistophora*. RP Yanong

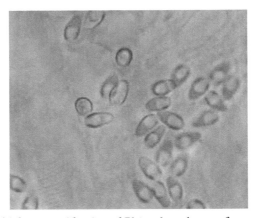

Figure 11.18B A higher magnification of *Pleistophora* that confirms these are a microsporidian due to the single posterior vacuole giving microsporidia a "wooden shoe" appearance. RP Yanong

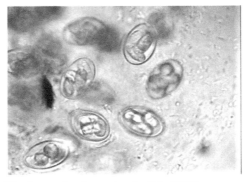

Figure 11.19 Coccidia from a fecal sample of a fish. RP Yanong

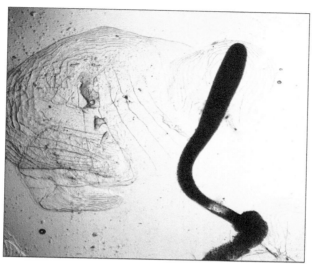

Figure 11.20A Skin scrape with worm-like structure in cardinal tetra (*Paracheirodon axelrodi*) with *Dermocystidium*. RP Yanong

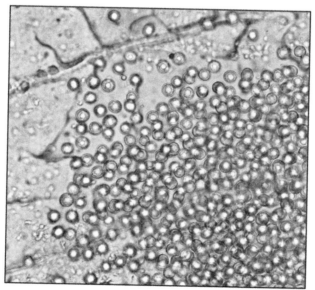

Figure 11.20B Close up of wet mount squash of worm-like structure demonstrating numerous circular *Dermocystidium* spores, each with a refractile body. RP Yanong

CHAPTER 11: INFECTIOUS DISEASES

Figure 11.21A   Pacific herring with *Ichthyophonus* infection. Note dark nodules (spores of the parasite) on skin, contributing to "sandpaper" texture. P Hershberger, US Geological Survey

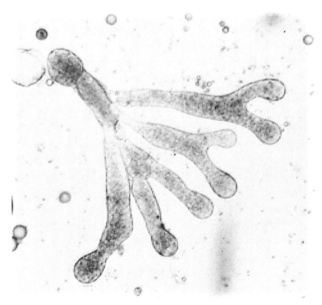

Figure 11.21B   Wet mount/post-mortem germination of *Ichthyophonus* spores. Note "hyphae-like" structures. P Hershberger, US Geological Survey

Table 11.5 Parasites: Metazoan - Worm/Worm-Like

CATEGORY	NAME	CLINICAL SIGNS	FEATURES	LIFE CYCLE (D OR I)	MANAGEMENT
Platyhelminthes	Cestode/Tapeworm		May be present in various tissues as metacestode (plerocercoid – may see encysted stage with scolex/four suckers in some species) or adult (flat, segmented) in GI tract	I	If adult in GI tract, praziquantel
	Monogenean – Gyrodactylid (F, S, Br); *Gyrodactylus* and others	Lethargy, increased mucus, anorexia, clamped fins, flashing, increased respiration, skin/gill/eye damage; primarily external, but some species may be found within internal organs, e.g., the stomach	"Livebearing" – adult with hooks, and hooks of internal embryo visible; no eye spots; V-shaped head; ~0.5–1.1mm length	D – livebearing; may have three generations in one worm	Formalin, praziquantel, organophosphates (fish species sensitivities), saltwater dips (FW fish), FW dips (SW fish); egg-producing species: must treat system for duration of life cycle; livebearing species, e.g., *Gyrodactylus* – shorter treatment (no egg stage); prolonged hyposalinity (gradual drop to 16-20 ppt, then held for 1-4 weeks or more) has been effective against capsalids but lower range salinity is less tolerated and for shorter duration for more stenohaline marine fish spp.
	Monogenean – Dactylogyrid (*Dactylogyrus* and others), Ancyrocephalid (F, S, Br)		Eye spots ("egg laying"); ~0.3–1.6mm length	D – eggs; life cycle 9–15 days ~25°C (77°F)	
	Monogenean –Capsalid (*Neobenedenia*, *Benedenia*, others) (S, Br)		Larger than dactylogyrid/an cyrocephalid/gyrodactylid, anterior "suckers," rounded haptor; ~2–9 mm length	D-eggs, brown, tetrahedral, with sticky filaments; life cycle ~2–3 weeks ~25°C (77°F)	
	Monogenean –Polyopisthocotylean (F, S, Br)		"Clamps" used to attach/ macerate gills for nutrition; blood feeders; ~4–33mm length	D – eggs	
	Digenean	Skin, fin, muscle may appear as white or black cyst/small mass/"grub"; gill, may result in structural damage and loss of respiratory tissue	Encysted stage (metacercariae) often seen on or in fish; adult stage with visible oral and ventral suckers, less common, but may be found in e.g., GI tract; 0.185–0.035mm	I – FW tropical species, often require bird, snail, fish hosts; SW species – snails, other hosts	Control snails, birds in aquaculture; potentially surgical removal in individual pet fish; praziquantel may kill metacercariae but then incite inflammatory response; if adults in GI tract, praziquantel

Table 11.5 (Continued)

CATEGORY	NAME	CLINICAL SIGNS	FEATURES	LIFE CYCLE (D OR I)	MANAGEMENT
	Turbellarian (S)	"Black ich" "Black grub" appearance (similar to encysted digenean), or epithelial proliferation	75um–450um; shorter body length relative to body diameter (compared with nematodes); eye spots and cilia; no hooks	D – life cycle "similar" to "ich": stages on and off the fish, duration ~ 10 days (24.5°C (76 °F); one adult – up to 160 juveniles	Formalin, organophosphates; FW dip may be necessary to reduce load.
Nematode	Capillarid spp. (e.g., Capillaria) F, Br, S	Intestinal worm; mature females contain barrel-shaped eggs with "plugs" on both sides	Males approx. 5.4–7.4 mm; gravid females approx. 9.4–16.5 mm	D – can spread from fish to fish and autoinfect	Fenbendazole, levamisole
	Eustrongylides	Grossly visible; coiled, long, red worm in coelom	I – larval stage found in fish; 11–83 mm; often multiple worms in coelom of one fish	I – fish intermediate host, wading bird final host; oligochaete worm intermediate host in some spp.	Prevention – control bird interactions and oligochaete worms; surgical removal if necessary for display or aquarium/pet fish
Arthropod	Pentastomid	Worm-like "grub"/mass/cyst on skin/muscle, distended coelom	Nymph stage (3–10 mm TL) in fish is relatively short, fat, and coiled; can be external or in various tissues internally; obvious segmentation (annuli) of the body; small hooks on anterior end; coiled nymph stage	I – final host, aquatic reptiles (turtles, snakes, alligators)	Control aquatic reptiles; no treatment for nymph stage in fish; surgical removal if logistically viable for display or aquarium specimens
Annelid Hirudinacea	Leech	Visible, attached, segmented worm on body, fins, operculum, oral cavity; reddened areas, erosion, ulceration from sucker attachment	Annelid – segmented worm, with anterior and posterior suckers	D – lay cocoons	Organophosphates most effective or in light infestations manual removal of worms and cocoons

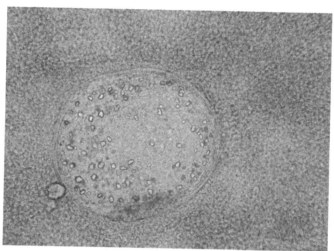

Figure 11.22A Cestode larva in the spleen of a fish. RP Yanong

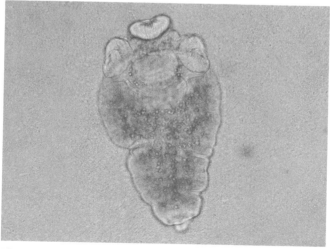

Figure 11.22B A section of a cestode from the GI tract of a black ghost knifefish (*Apteronotus albifrons*). Note suckers on the scolex. M Hyatt/J Meegan/RP Yanong

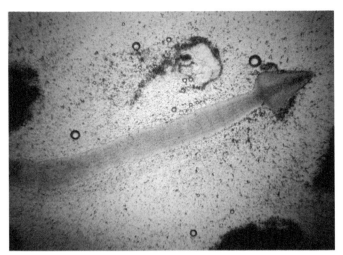

Figure 11.22C Adult Asian tapeworm (*Bothriocephalus acheilognathi*). Note the heart-shaped scolex. D Pouder.

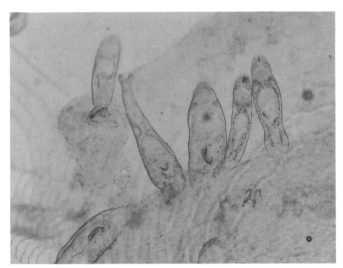

Figure 11.23A A large number of the livebearing monogenean *Gyrodactylus* from a skin biopsy. Note the V-shaped head, lack of eye spots, and internal ovoid embryo with hooks. RP Yanong

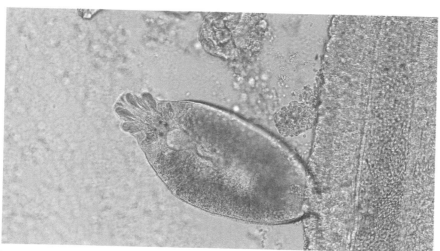

Figure 11.23B A single egg-laying dactylogyrid monogenean on a gill biopsy sample. Note scallop-shaped head and eye spots. RP Yanong

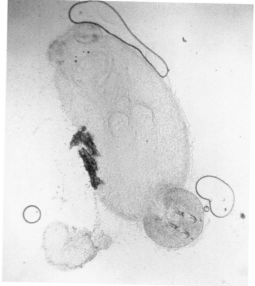

Figure 11.23C Microscopic view of the egg-laying capsalid monogenean, *Neobenedenia* sp. Note anterior suckers and eye spots and round posterior opisthaptor with anchors. RP Yanong

Figure 11.23D A *Neobenedenia* sp. monogenean after removal from a marine fish exposed to a freshwater dip. Note darkened internal features. RP Yanong

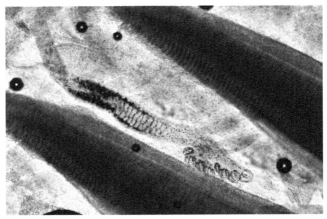

Figure 11.23E Polyopisthocotylean (monogene) infesting the gills of a fish. UF IFAS Tropical Aquaculture Laboratory

Figure 11.23F   Polyopisthocotylean (close up). Note "clamp-like" haptor used to attach and grind gill tissue. UF IFAS Tropical Aquaculture Laboratory

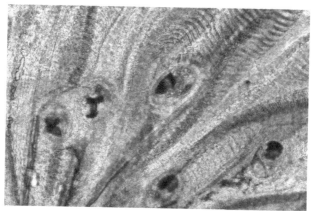

Figure 11.24   Digenean metacercaria within gill tissue. RP Yanong

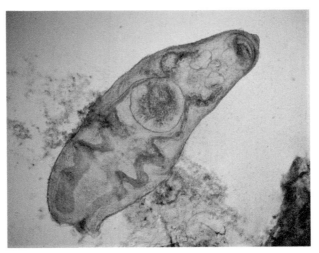

Figure 11.25 A mature digenean trematode from the GI tract of a spiny eel (Mastacembelidae). Note the anterior oral and more central ventral sucker for attachment and paired reproductive tract. M Hyatt/J Meegan/RP Yanong

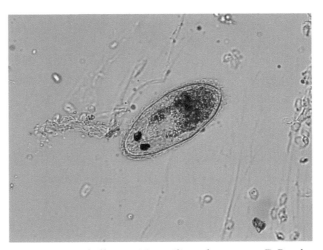

Figure 11.26 A parasitic turbellarian. Note cilia and eye spots. D Pouder

Parasitic Diseases Of Aquarium Fish

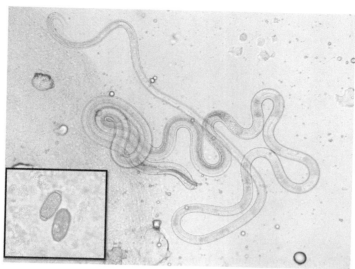

Figure 11.27A Nematode parasite, *Capillaria pterophylli*, from the intestine of an angelfish (*Pterophyllum scalare*). Note bi-operculate eggs (inset). M Hyatt/J Meegan/RP Yanong

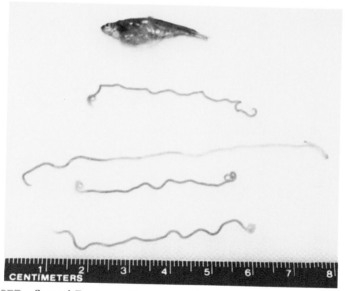

Figure 11.27B Several *Eustrongylides* (nematode) worms from the coelomic cavity of a zebra danio (*D. rerio*). These resemble red "spaghetti," are relatively large, and typically coiled. DB Pouder

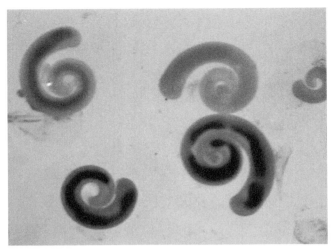

Figure 11.27C   Pentastome nymphs removed from within the coelom of a swordtail. RP Yanong

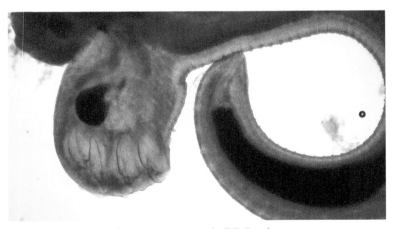

Figure 11.27D   Close up of Pentastome nymph. DB Pouder

Figure 11.28A Leeches on the barbels of a channel catfish (*Ictalurus punctatus*). GA Lewbart

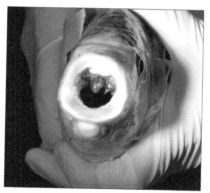

Figure 11.28B Leeches in the mouth of a tilapia. RP Yanong

Figure 11.28C Close up of leeches in the mouth of a tilapia. RP Yanong

CHAPTER 11: INFECTIOUS DISEASES

Table 11.6 Parasites: Metazoan - Crustacea and Myxozoa

CATEGORY	NAME, F, S, BR	CLINICAL SIGNS	MOVEMENT/ FEATURES	LIFE CYCLE (D OR I)	MANAGEMENT	NOTES
Crustacea: Branchiura (F, Br)	*Argulus* (fish louse) F, S, Br (*A foliaceus*: 3–7 mm x 2–4 mm)	Lethargy, flashing, isolation, anorexia, visible parasites –skin, gill chamber, mouth	Can detach, survive several days off host; dorsoventral (DV) flattened, ovoid to round, eye spots and two prominent "suckers"	D – 30–60 days (temperature, spp. dependent); numerous life stages; juveniles can overwinter in mucus; continue to molt after maturity; lays eggs on substrate	Organophosphates, chitin/molt inhibitors (e.g., Dimilin® (diflubenzuron) lufenuron), emamectin benzoate (water or oral); legality issues for use of some drugs/pesticides	May serve as vector for virus, bacteria; males and females parasitic
Crustacea: Copepod (F, S, Br)	*Lernaea* (anchorworm) F (3–12 mm length)	Visible "thread" like females in skin, fins, gills, oral cavity; respiratory distress, lethargy, hemorrhages	Female "anchors" into fish, egg sacs attached to female	D – 18–25 days, 29°C (84°F); numerous life stages; temperatures < 20°C (68°F), juveniles unable to complete development; at temperatures < 14°C (57°F), females unable to reproduce, but can overwinter	Similar to *Argulus* above; also, juvenile *Lernaea* more susceptible to hypersalinity, 4.8g/L x 30 days may help break life cycle; can physically remove female with eggs but anchor may remain	Attachment may result in secondary infections and inflammation
	Ergasilus (> 2mm) F (Br, S)	Visible as small "spots" on gills	Female anchors with grasping antennae to gills, eggs sacs attached to female	D – numerous life stages	Similar to *Argulus* above; physical removal	—
	Salmincola F (female 1.6–3.7 mm without egg case)	Visible on skin, fin, gills; gill damage results in respiratory distress	Opaque, white female attaches via conical bulla; paired egg sacs often seen	D	Similar to *Argulus* above	One of several species of parasitic copepod known as the "gill louse" or "gill maggot" (e.g., *Ergasilus*); similar general morphology

Table 11.6 (Continued)

CATEGORY	NAME, F, S, BR	CLINICAL SIGNS	MOVEMENT/ FEATURES	LIFE CYCLE (D OR I)	MANAGEMENT	NOTES
Crustacea: Isopod (S, F, Br)	Cymothoidae S (some F/Br)	Visible and attached to skin, gills/gill cavity, mouth, or burrowed into fish in pouch	Dorso-ventrally flattened, segmented exoskeleton, some species	D – varies; female primary parasite	Some observations of cleaner fish removing juvenile stages; organophosphates and/or manual removal	Cymothoidae: obligate fish parasite; protandrous hermaphrodites
Cnidaria: Myxozoa (F, S, Br)	Henneguya, Myxobolus, others	Varies, depending on site of infection	Non-motile; microscopic spores in myxospore cysts; spores can vary in shape and size by species; 1–15 polar capsules	Most species, I – requires invertebrate final host; few species, D (e.g., some GI myxozoa)	No known treatment; if external, surgical removal; manage life cycle	Can be incidental finding, depending on severity; classified as parasitic Cnidaria

(Hoffman, 1999; Landshaw and Feist, 2001; Noga, 2010; Stoskopf, 1993; Woo, 2006; Woo and Bruno, 2011)

Figure 11.29 The branchiuran crustacean parasite, *Argulus* sp., on a koi (*Cyprinus rubrofuscus*). D Pouder/RP Yanong

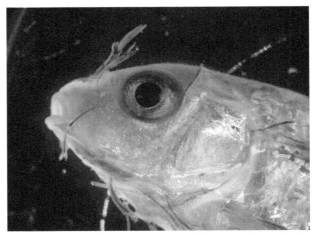

Figure 11.30A Numerous anchorworm parasites (*Lernaea* sp.) on a koi. UF IFAS

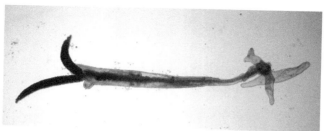

Figure 11.30B Anchorworm (*Lernaea* sp.) adult female removed from koi. Note paired egg sacs. UF IFAS

Figure 11.31B Close up of *Ergasilus* sp. attached to the gill of a crappie. F Meyer

Figure 11.32 Several *Salmincola* sp. crustaceans on a rainbow trout (*Oncorhynchus mykiss*). GA Lewbart

CHAPTER 11: INFECTIOUS DISEASES

Figure 11.33 An isopod on the body of a wild Spanish mackerel (*Scomberomorus maculatus*). GA Lewbart

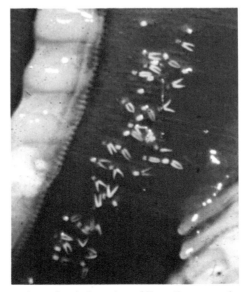

Figure 11.33A Gross view of an infestation of *Ergasilus* sp. on the gills of a catfish. F Meyer

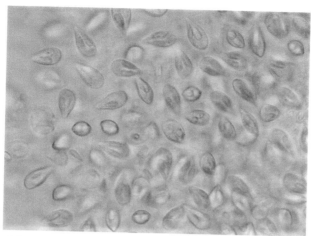

Figure 11.34 Myxozoan (close up) from a gill lesion in a goldfish. RP Yanong

BACTERIAL DISEASES OF AQUARIUM FISH

Bacterial diseases can be the primary causes of disease but are often secondary to trauma, poor water quality, parasites, or other stressors. Although most require culture and sensitivity to facilitate identification and optimize drug choice, some, such as *Flavobacterium columnare* and related species, as well as epitheliocystis-causing bacteria, can be tentatively identified using wet mount biopsies, and managed appropriately.

General Clinical Signs
- Non-specific, but can include inappetence, changes in behavior (abnormal attitude in the water, lethargy, spinning, abnormal swimming), and external lesions (fin or body erosions, ulcerations, hemorrhages, coelomic distension, edema, darkening, exophthalmia).

Tables **11.7** and **11.8** provide general information on important groups of bacterial pathogens and bacterial diseases of fish. For treatable bacterial diseases, culture and sensitivity are preferred, but management options are suggested.

Table 11.7 Common Bacterial Diseases of Aquarium Fish

COMMON BACTERIAL DISEASE GROUPS	REPRESENTATIVE GENERA/DISEASE	CLINICAL SIGNS	TARGET SPECIES	DIAGNOSTICS	MANAGEMENT	ADDITIONAL NOTES
Gram Negative, Systemic/Ulcerative	*Aeromonas*, *Vibrio*, *Pseudomonas*, *Edwardsiella*, *Photobacterium*	Hemorrhages, erosion, ulceration, ascites, lethargy, anorexia, abnormal behavior, mortalities	Wide host range	Culture/ID (kidney, ulcer lesions, blood); atypical aeromonas salmonicida – PCR	Improve husbandry; antibiotics based on culture and sensitivity	Atypical *A. salmonicida* is a common cause of koi/goldfish ulcer disease and can be difficult to culture; qPCR may be required for ID
Gram Negative, Primarily External	*Flavobacterium* spp. ("columnaris disease" – *F. columnare* and other related species, and a few other genera), bacterial gill disease (BGD), *Tenacibaculum* (SW) and other related bacteria	Erosion, "saddleback lesions" (white or paler area around dorsal fin); whitish areas and/or fraying or ulceration of skin, fins; necrosis/mottling of gills; mortalities	Wide host range; *Tenacibaculum* for marine species	Skin scrape: wet mount – columnaris: "flexing bacterial haystacks"; culture/ID require special media, PCR; other bacteria also may be seen	Improve husbandry; immersion treatments ($KMnO_4$, diquat dibromide, antibiotics); oral antibiotics may be warranted	Some infections can go systemic
Acid Fast	*Mycobacterium* spp., *Nocardia* spp.	Non-specific, lethargy, wasting/cachexia, ascites, ulcers, abnormal pigmentation, hemorrhage	Wide host range	Necropsy – granulomas – kidney, spleen, other tissues, typical but granulomas not seen in some cases; AFB stain, culture/ID;	No effective treatment; some drugs may appear to "prolong" life, but no good data; supportive care if euthanasia; euthanasia not an option; zoonotic	Environmental bacteria, can be difficult to kill in the environment using standard disinfection methods; immunocompromised, stressed fish more susceptible; clinician must consider the contagious nature and zoonotic potential of this pathogen.

Figure 11.35A Deep ulceration and hemorrhages in koi (*Cyprinus rubrofuscus*) with atypical *Aeromonas salmonicida* infection. RP Yanong

Figure 11.35B Smaller ulcers and hemorrhagic lesions in koi (*Cyprinus rubrofuscus*) with atypical *Aeromonas salmonicida* infection. RP Yanong

Figure 11.36A Columnaris disease in female swordtails. Note pale/whitish "saddleback" lesion around dorsal fin of bottom fish and erosion and pale/whitish lesions around caudal peduncle of middle fish. RP Yanong

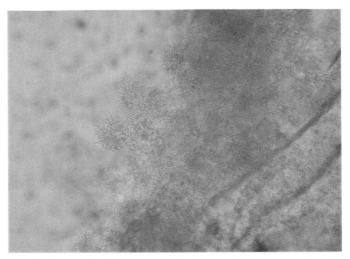

Figure 11.36B Micrograph of wet mount of columnaris bacteria on fin of fish. Note "haystack-like" appearance of aggregates of long, flexing bacteria. RP Yanong

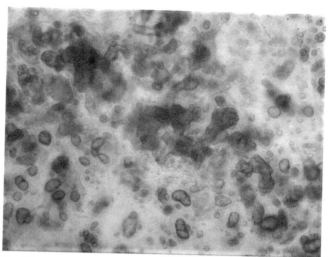

Figure 11.37A   Numerous granulomas in the posterior kidney of a cichlid with mycobacteriosis. Note "rock-like" appearance and, upon closer inspection, the presence of an "onion-skin" layering around these granulomas caused by epithelioid macrophages that surround the pathogen. Granulomas are not specific to mycobacteriosis but are often associated with it. Additional testing (special stains, culture, qPCR) is required for confirmation. D Pouder

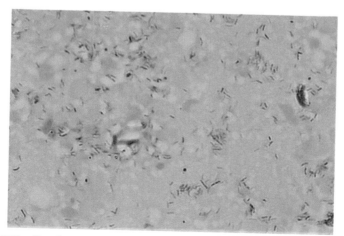

Figure 11.37B   Benchtop acid fast stain of dried tissue wet mount demonstrating numerous "banded" acid fast bacteria consistent with mycobacteria infection. D Pouder

Table 11.8 Other Important Bacterial Diseases of Aquarium Fish

OTHER IMPORTANT BACTERIAL DISEASES	TYPICAL CLINICAL SIGNS	TARGET SPECIES	DIAGNOSTICS	MANAGEMENT	ADDITIONAL NOTES
Streptococcus/Lactococcus (Gram Positive Cocci)	Hemorrhages, exophthalmia, neurologic (e.g., spinning/abnormal swimming), rapid mortalities	Wide host range	Culture, ID, and sensitivity of blood, brain, or kidney/spleen; gram stain of brain impression smear or blood smear reveal gram positive cocci	Oral erythromycin, oral or injectable florfenicol	Can spread quickly through water
Erysipelothrix piscisicarius (Gram Positive Rod)	Necrosis and ulceration of mouth, mortalities	Tetras (characids), barbs (cyprinids)	Culture and ID; may need to work with advanced fish disease/ microbiological diagnostic laboratory	Currently difficult to treat, but attempt based on culture and sensitivity; autogenous vaccines have proven effective in the aquaculture industry	Cull individuals showing gross clinical signs – these often do not respond to treatment
Francisella orientalis (Gram Negative Cocco-Bacilli)	Lethargy, anorexia, darkening, crowding, abnormal swimming, exophthalmia, mottled gills; may have no signs other than acute mortalities	Wide host range, (warm-water fish)	Necropsy – nodules/granulomas – kidney, spleen, other tissues; AFB negative; culture (special media)/ID (qPCR)	Oxytetracycline or florfenicol; can be difficult to treat depending on stage of disease; culture and sensitivity require special media	Gross and general histopathology may appear similar to mycobacteriosis
Epitheliocystis (Chlamydiales, Gamma-Proteobacteria, or Beta-Proteobacteria)	Small white nodules/cysts in gills or skin, respiratory distress, anorexia, increased mucus, weakness, lethargy	Wide host range, FW/SW	Wet mount – small "granular" cysts in gills; histopathology: enlarged epithelial cells with basophilic, encapsulated, bacterial filled spherical cysts	May respond to immersion treatment with antibiotics (oxytetracyline, or others); culling may be required in severe cases	– not culturable; poorly studied; PCR ID

Figure 11.38 Zebra danio (*D. rerio*) with streptococcosis. Note exophthalmia and external hemorrhages. D Pouder.

Figure 11.39 Tiger barbs (*Puntius tetrazona*) with *Erysipelothrix* infection have severe ulceration, inflammation, and necrosis of the mouth/face. RP Yanong

CHAPTER 11: INFECTIOUS DISEASES

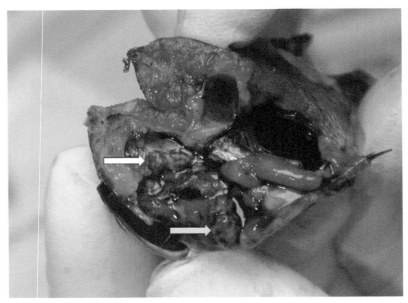

Figure 11.40 Juvenile tilapia (*Oreochromis* sp.) cut transversely through the center of the fish. The kidney (white arrow) is greatly enlarged with pale white nodules, and the spleen (yellow arrow) is also enlarged and nodular. RP Yanong

Figure 11.41A Small cystic structures seen in gill wet mount of an oscar with epitheliocystis. RP Yanong and G Trende

Figure 11.41B Close up of oscar (*Astronotus ocellatus*) gill wet mount showing granularity of cystic structures in epitheliocystis. RP Yanong and G Trende

VIRAL DISEASES OF AQUARIUM FISH

Although viruses are important to rule out – especially when morbidity and mortality are high and rapid – other infectious and non-infectious factors, including water quality, toxins, or system issues should also be considered. A few viral diseases of aquarium fish have distinct lesions (e.g., koi pox, lymphocystis) that are often presumptively identified and managed on gross appearance or microscopic wet mount alone. However, other viral diseases have more non-specific signs that share similarities with other infectious and some non-infectious diseases, so virology, including polymerase chain reaction (PCR), transmission electron microscopy (TEM), or virus isolation by cell culture, in addition to supportive light histopathology, will be critical for proper diagnosis (see **Tables 11.9** and **11.10** and **Figures 11.42–11.47B**).

Finally, because hundreds of species comprise the aquarium hobby and public aquaria, it is expected that "new" emerging viral diseases with varying species susceptibilities will be identified in the future.

Table 11.9 **Viral Diseases of Koi and Goldfish**

DISEASE	VIRAL TAXONOMY	TARGET SPECIES	COMMON CLINICAL SIGNS	DIAGNOSTICS	MANAGEMENT	COMMENTS
Koi (carp) pox	Cyprinid herpesvirus 1	Koi/common carp	"Dripped candle wax" lesions on skin, fins	Spring (<68°F/20°C); Clinical signs/lesions; histopath	In older juveniles/adults, resolves when temperatures increase	May recrudesce seasonally; purportedly can be problematic with young fish
Herpesviral hematopoietic necrosis (HVHN)/Goldfish hematopoietic necrosis virus disease (GFHNVD)	Cyprinid herpesvirus 2	Goldfish	Lethargy, at bottom, increased respiration	Temperature 59°–68°F; Histopath (head/trunk kidney, spleen necrosis), PCR	No treatment; palliative; survivors most likely carriers	Endemic on many commercial farms; disease triggered in stressed naive fish exposed to carriers; increasing temps to 27 degrees C may reduce losses
Koi herpesvirus disease (KHVD)	Cyprinid herpesvirus 3	Koi/common carp; goldfish (and some other species) can be inapparent carriers	Non-specific, mottled red/white gills, bleeding gills, sunken eyes, pale patches on skin, notched nose	Temperature more commonly 70°–82°F (possibly below 70°F); virus isolation, PCR, serology	No treatment; because survivors may become carriers, introduction of new (naive) koi problematic	USDA/WOAH reportable disease

Table 11.9 (Continued)

DISEASE	VIRAL TAXONOMY	TARGET SPECIES	COMMON CLINICAL SIGNS	DIAGNOSTICS	MANAGEMENT	COMMENTS
Spring viremia of carp (SVC)	Rhabdoviridae; Rhabdovirus carpio	Koi/common carp, goldfish, other large Asian carps	Lethargic, abnormal swimming, darkened, ascites, exophthalmos, hemorrhages, anemia/pale gills	Temperature (50°–64°F); virus isolation, PCR	Depopulation recommended; work with state and federal veterinarians	USDA/WOAH reportable disease; temperature <18°C/springtime;
Carp edema virus disease (CEV)/koi sleepy disease	Poxviridae	Koi/common carp	Fish lie motionless on the bottom, unless disturbed then swim briefly before returning to bottom; skin erosion, hemorrhage; edema; swollen gills	Temperature 59°–77°F; clinical signs (sleepy), swollen gills/clubbing, PCR	0.5% salt "may" reduce morbidity/mortality	15°–25°C in many cases, but some strains cause disease at lower temperatures

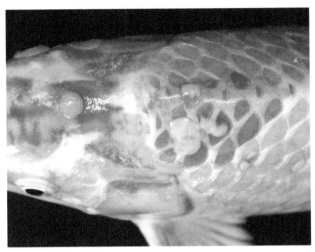

Figure 11.42 "Candlewax" lesions seen on the dorsum of a koi (*Cyprinus rubrofuscus*) with koi pox. A Goodwin, USFWS

Figure 11.43 Mottled red and white (necrotic) gills (operculum removed) and endophthalmia seen in koi (*Cyprinus rubrofuscus*) with koi herpesvirus disease. D Pouder

Figure 11.44 Mirror carp (*Cyprinus rubrofuscus*) with exophthalmia and external hemorrhages resulting from SVC infection. A Goodwin, USFWS

Figure 11.45A Koi (*Cyprinus rubrofuscus*) with carp edema virus (CEV) infection, aka koi sleepy disease, infection lying on their sides on the bottom of a tank. If the sides of the tank are struck, the fish will swim up briefly before heading back down. J Shelley

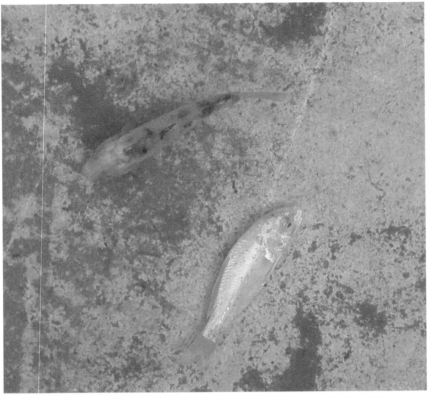

Figure 11.45B Close up of another case of CEV. Koi (*Cyprinus rubrofuscus*) lying on its side will respond to stimuli and swim up briefly and then return to the bottom. G Trende

Table 11.10 Viral Diseases of Tropical Aquarium Fish

OTHER TROPICAL FISH VIRUSES/ VIRAL DISEASES	VIRAL TAXONOMY	TARGET SPECIES	COMMON CLINICAL SIGNS	DIAGNOSTICS	MANAGEMENT	COMMENTS
Viral nervous necrosis	Betanodavirus	Primarily marine tropical fish	Neurologic or visual impairment, as CNS is affected; swim bladder hyperinflation in fry	Histovacuolation in CNS tissues (but beware of artifactual "holes" from poor fixation); quantitative polymerase chain reaction (qPCR ID)	Dx/ID is sacrificial; no treatment for affected fish	—
Megalocytivirus	Iridoviridae	Wide host range, freshwater (FW) to saltwater (SW) species	Non-specific	Histobasophilic "megalocytes" (enlarged cells) in affected tissues; qPCR ID confirmatory	No treatment; commercial vaccines available for food fish internationally but currently none for aquarium fish in N. America	Red seabream iridovirus is USDA/ WOAH reportable
Lymphocystivirus/ Lymphocystis	Iridoviridae	Higher orders of teleosts; NOT seen in cyprinids, catfishes, other lower orders	Wart-like nodules on skin, fin, gills; can be internal	Wet mount (look like balloons) or histo; enlarged, viral laden fibroblasts	Dependent on degree of infection and location of nodules; self-limiting in minor cases; symptomatic treatment, removal of nodules to reduce infectious load	—

(Continued)

Table 11.10 (Continued)

OTHER TROPICAL FISH VIRUSES/ VIRAL DISEASES	VIRAL TAXONOMY	TARGET SPECIES	COMMON CLINICAL SIGNS	DIAGNOSTICS	MANAGEMENT	COMMENTS
Picornavirus	Picornaviridae	Emerging	Associated with lethargy, color changes, inflammation, and erythema, hemorrhages, exophthalmia, abnormal position	Histopathology – individual cell necrosis, mononuclear cell inflammation in branchial cavity, pharynx, esophagus and/or stomach; possible large basophilic inclusions in pharyngeal mucosa; PCR ID	—	Emerging group; associated with disease in clownfish; asymptomatic cases seen in carp, zebrafish, bluegill and several others

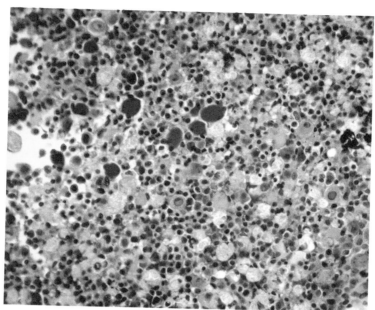

Figure 11.46 Histopathology of spleen infected with *Megalocytivirus*. Typical findings include dark basophilic inclusions in enlarged cells ("megalocytes") that are often associated with necrosis in the spleen and kidney but may be found in other organs without it. RP Yanong

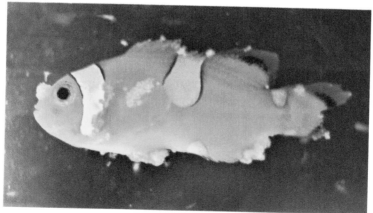

Figure 11.47A Clownfish (*Amphiprion ocellaris*) with severe lymphocystis infection. Note "wart-like" nodules distributed throughout body and fins. RP Yanong

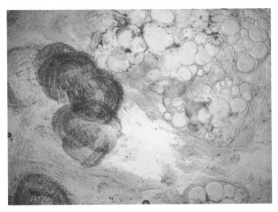

Figure 11.47B Wet mount of skin scrape from clownfish (*Amphiprion ocellaris*) with lymphocystis. Round to ovoid cells of varying sizes in "grape-like" clusters are greatly enlarged fibroblasts filled with viral particles. RP Yanong

FUNGAL DISEASES OF FISH

The group of fish pathogens called "fungi" are a taxonomically varied set of organisms that share a similar saprophytic lifestyle, albeit with different biological characteristics. The water molds, including *Saprolegnia*, a common pathogen of freshwater fish, are among the better known. Most are considered secondary invaders relying on trauma or immunosuppressive conditions to cause disease. Some are considered more primary in their pathogenicity. Regardless, many can become aggressive invaders with no effective chemotherapeutic options. A few better-known examples are provided, but positive, confirmatory identification, as for other pathogens above, require working with a specialized fish disease laboratory (see **Table 11.11** and **Figures 11.48A–11.50B**).

Fish can be infected by a variety of fungi and fungi-like saprophytic and parasitic spore-producing organisms that include the true fungi and water molds, among others. Infection by this group is often secondary to suboptimal water quality and environmental conditions (including temperature), bacterial or parasitic infections, trauma, and/or other handling stressors that impair or erode mucus and skin integrity and defenses and also result in immunosuppression.

(*Note: some organisms considered closely related to fungi or fungi-like share protist-like traits (Microsporidia and Mesomycetozoea) and are included in **Table 11.4**).

Table 11.11 Fungal Diseases

DISEASE	TAXONOMY	TARGET SPECIES (F- FRESH, S- SALTWATER, BR - BRACKISH)	COMMON CLINICAL SIGNS	DIAGNOSTICS	MANAGEMENT	COMMENTS
Saprolegniasis	Oomycete- *Saprolegnia*	F, Br	Cottony white, gray, brown, red, green mat on fin, body, gills; may be accompanied by hemorrhages	Wet mount, tentative Dx – presence of aseptate hyphae and "cattail" like zoosporangia (spore case); culture, morphologic and qPCR ID	Can be a challenge to manage, but easier if identified early and if only on the fins; 3–5 ppt salt to help with osmoregulatory function; formalin, bronopol, malachite green; increasing temperature may help	Zoospores released from the spore case, are motile, and may be confused with other parasites; infection results in severe osmoregulatory dysfunction
Epizootic ulcerative syndrome	Oomycete- *Aphanomyces invadans*	F, Br	Severe, deep ulceration, and granulomatous inflammation and necrosis of the skin/ muscle; can invade body cavity/internal organs	Wet mount, tentative Dx – presence of hyphae with "grape-like" zoosporangia, located within deep ulcerative lesions of skin/muscle; granulomas in muscle/ skin; culture, morphologic and qPCR ID	Difficult to treat, especially if severely affected; malachite green, formalin; increased salinity 10–20 ppt for 1 hour; 3–5 ppt salt for osmoregulatory aid	WOAH, USDA reportable; found worldwide; endemic in wild fishes in mid-Atlantic to southeast coastal areas of the US

(Continued)

Table 11.11 (Continued)

DISEASE	TAXONOMY	TARGET SPECIES (F-FRESH, S-SALTWATER, BR-BRACKISH)	COMMON CLINICAL SIGNS	DIAGNOSTICS	MANAGEMENT	COMMENTS
Fusariomycosis	Eumycota – *Fusarium* spp.	F, S, Br; many species can be affected, but especially Pomacanthidae (marine angelfishes) and Scaridae (parrotfishes), and Sphyrinidae (hammerhead/bonnet head sharks)	Angelfish – raised pustules around lateral line system, often with invasion of underlying muscle/bone; can go systemic; parrotfish – localized epidermal lesions (skin defects, ulcers, lepidorthosis, necrotizing dermatitis, myositis, perforation to coelom, infection of internal organs, death); sharks – myositis, myonecrosis, papules with purulent exudate on dorsal and ventral surface of head and lateral canals	Culture, morphologic and qPCR ID	Chemotherapeutics ineffective; often temperature issue; try to mimic natural thermal cycles	Environmental – common on plants and in tropical/subtropical soils

(Yanong, 2003)

Fungal Diseases Of Fish

Figure 11.48A African cichlid with saprolegniasis has numerous "cottony" tan to yellowish masses spread throughout the body. Some of these are associated with hemorrhages. RP Yanong

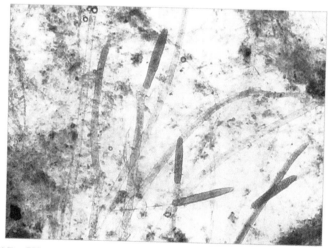

Figure 11.48B Wet mount of *Saprolegnia* with "cattail-like" zoosporangia. RP Yanong

Figure 11.49A Grey mullet (*Mugil cephalus*) with severe ulceration, inflammation, and necrosis caused by *Aphanomyces invadans*. Photo courtesy of FWC Fish and Wildlife Research Institute

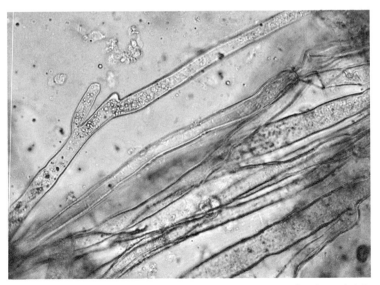

Figure 11.49B Wet mount micrograph of thin-walled, aseptate hyphae of *Aphanomyces invadans*. Photo courtesy of FWC Fish and Wildlife Research Institute

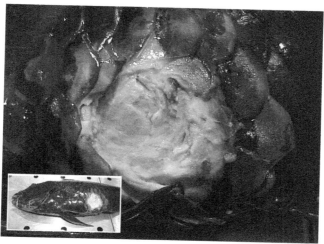

Figure 11.50A   Necrotic ulcerative skin lesions in midnight parrotfish (*Scarus coelestinus*) with fusariomycosis. S Terrell

Figure 11.50B   Reddened and hemorrhagic lesions on ventral surface of head of bonnethead (*Sphyrna tiburo*) caused by fusariomycosis. S Terrell

REPORTABLE DISEASES

Reportable diseases of aquarium fish are those diseases that are of concern and for which, if suspected and/or confirmed, reporting to the relevant federal and/or state animal health officials is required.

Internationally reportable diseases of fish are listed by the World Organization for Animal Health (WOAH) (woah.org), formerly known as the

Office International des Epizooties (OIE). The two most important aquarium fish diseases listed include koi herpesvirus (KHV) and spring viremia of carp (SVC). Two other listed diseases seen less commonly include Red Sea bream iridoviral disease (RSIVD) and epizootic ulcerative syndrome (EUS) caused by *Aphanomyces invadans*, a water mold relative of *Saprolegnia* that has been identified in wild fish in the US.

The US Department of Agriculture (USDA) Animal & Plant Health Inspection Service (APHIS) National List of Reportable Animal Diseases (NLRAD) (https://www.aphis.usda.gov/aphis/ourfocus/animalhealth/monitoring-and-surveillance/nlrad/ct_national_list_reportable_animal_diseases) includes EUS, RSIVD, and SVC as notifiable diseases, and KHV as monitored (endemic in the US but periodic updates to the WOAH).

Other diseases may be reportable locally; contact the state's animal health office for more information.

ZOONOSES

- Zoonoses reported from aquarium fish, although possible, are relatively uncommon, especially given the large number of aquariums and aquarium fish maintained nationally and globally.
- People who have compromised or suboptimal immune systems due to age, underlying diseases, or use of immunosuppressive medications will be much more susceptible to these and other infectious agents. These people should follow extra caution and avoid direct exposure to fish or aquatic systems.
- Regardless, emphasis on good personal hygiene and biosecurity, aquarium sanitation and equipment disinfection will significantly reduce any risks.
  - Good hygiene practices include washing hands or other exposed areas after working in aquarium or pond systems or handling fish.
  - "Mouth" siphoning (i.e., starting the suction of a siphon hose using one's mouth) should be avoided.
  - Avoid any contact with water or fish if abrasions, cuts, or other skin lesions are present on hands or arms (or other areas, including feet and legs, if, for example stepping into ponds), or, if contact is unavoidable, reduce risk by use of personal protective equipment, e.g., arm length gloves, etc., followed by good hygiene practices.
  - Good system management, sanitation, disinfection, and biosecurity practices will also help reduce the spread of zoonoses significantly.
  - Understand the fish species, have proper equipment (e.g., the right size and number of nets), and use proper handling techniques to minimize stress on the animal and the handler and to avoid puncture wounds, scrapes, or even bites from some species.

- Although much rarer cases of zoonoses from other aquarium fish pathogens that are not listed here are possible, the more relevant aquarium fish zoonotic pathogens of note are bacterial and are listed below (Boylan, 2011; Lowry and Smith, 2007). Clients should contact their personal physicians if they suspect an aquarium fish zoonosis.
- Clinical signs identified in humans from some fish bacterial pathogens are listed below; primary predisposing factors included penetration of human skin by puncture with sharp fin rays while handling fish, other skin trauma allowing entry, or ingestion of water, and underlying immunosuppression:
  - *Aeromoniasis* – can cause localized wound swelling and gastroenteritis.
  - *Edwardsiella* – can cause necrotic skin lesions and gastroenteritis.
  - *Mycobacterium* – (known as "fish tank granuloma" or "fish handler's disease") can cause raised or granulomatous nodules/dermatitis in humans.
  - *Salmonella* – not associated with fish disease but may be present in aquarium water (especially with turtles).
  - *Streptococcus* – can cause cellulitis, systemic arthritis, endocarditis, and meningitis.
  - *Vibrio* – necrotizing fasciitis, edema, swelling of the puncture wound.

CHAPTER 12
COMMON MEDICAL CONDITIONS

FATTY LIVER DISEASE

- Fish fed methyltestosterone to enhance color; history of slowly fading color; gradual wasting 1–2 months after purchase.
- Fish overfed or fed high-fat diet.
  - Differential diagnosis for color fading: liver disease, stress, cold shock, hypoxia; young goldfish frequently lose black markings.
  - Treatment: change diet; prognosis varies.

NEOPLASIA

- Diagnosis: enlarged abdomen and/or growth; radiography; biopsy.
- Pigment cell tumors (melanomas, erythrophoromas, and iridophoromas) are seen in aquarium fish, including goldfish and koi.
- Many tumors in wild fish are skin tumors. Some include benign epidermal hyperplasia (may be viral in origin); papillomas; sarcomas.
- Gonadal tumors are also seen in koi and other species (see **Figures 12.1–12.3**).
- Surgical removal is possible.

BUOYANCY PROBLEMS (SWIM BLADDER "DISEASE")

- Affects fish with swim bladders and may affect some species/varieties (e.g., oranda goldfish with swim bladder morphology altered from wild type) more commonly. If there is air in the GI tract, due to altered swim bladder morphology and location compared to the wild type, fish are anatomically more unstable and more likely to "roll."

DOI: 10.1201/9781003057727-12

Figure 12.1 *Synodontis* catfish male with gonadal tumor. D Pouder

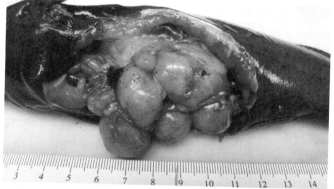

Figure 12.2 *Synodontis* catfish male – seminoma/mixed germ cell neoplasm masses. D Pouder

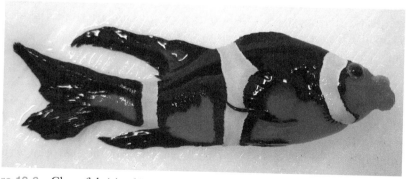

Figure 12.3 Clownfish (*Amphiprion ocellaris*) with an odontoma surrounding its mouth. RP Yanong

- Causes: supersaturation of water with air, infection/rupture of swim bladder, inner ear disease, GI disease (obstruction of pneumatic duct by food) or gas in the GI tract; congenital deformation.
- Treatment: treat inciting cause; aspirate air and/or surgical correction; ignore (some can eat upside down).
- Feed one to two green peas per day (unknown mechanism).
- Prevent by avoiding the use of dry food, or soak flakes/pellets before feeding (reduces air gulping).

SUPERFICIAL WOUNDS

- Caused by bacteria, parasites, trauma, aggression, or pre-spawning chase in ponds or tanks with rough rocks.
- Secondary bacterial infection may occur if the primary cause was not bacterial.
- Internal tissues of freshwater fish are hyperosmotic; internal tissues of salt-water fish are hypo-osmotic. Surface wounds disturb osmotic balance, may result in loss of salt/fluid balance and circulatory collapse.

SIGNS ASSOCIATED WITH WATER QUALITY PROBLEMS

- Stress of poor water quality can predispose fish to any disease.
- Excess mucus production, skin inflammation, gill erosion, hemorrhage.
- Pale gills (+/- telangectasia – "ballooning" of gill capillaries), skin erosions, gill lamellar fusion, and/or necrosis (noted at necropsy).
- Piping (gasping for air at water surface) indicates oxygen deprivation: increase aeration. This can also indicate high nitrite levels, which will impair oxygen uptake by red blood cells. Note: this can be normal behavior for partial air breathers (e.g., lungfish, eels); surface-dwelling fish (e.g., leaf fish); anabantoid (fish with labyrinth organs) bubble-nesters (fish that build nests for their eggs from bubbles they produce).
- Gas bubbles under the skin, fins, in the cornea/within the eye, or in various other tissues; exophthalmia; sudden death are seen in gas bubble disease. Note: not all cases of supersaturation include all of these clinical signs. Likewise, presence of bubbles in gill or skin vasculature may be iatrogenic or caused by trauma. However, if caused by supersaturation (excess nitrogen gas in the water) these may be due to:

- Moving fish from cold to warm water, defective pumps, defects in pipes, flow through systems using untreated source water that is supersaturated.
- Ponds: toxic runoff or acid rain pollution.

LATERAL LINE DEPIGMENTATION SYNDROME (LLD) OF MARINE FISH

- Formerly referred to as head and lateral line erosion (HLLE) syndrome.
- Acanthuridae (surgeonfishes, tangs) and Pomacanthidae (angelfishes), among others, are highly susceptible (see **Figure 12.4**).
- Superficial erosions of the head and face progressing down the lateral flank to and including the lateral line.
- Most likely numerous differing etiologies, most of which are unknown. One study demonstrated LLD/HLLE initiated by use of extruded coconut shell activated carbon filtration. Diet and nutrition are suspected as another cause, as some cases respond to vitamin C supplementation. Other suggested etiologies include poor water quality/chemistry, elevated redox potential, microbiome/bacterial imbalances, parasites, among others.
- Encouraging results with topical application of Regranex®, a human platelet growth factor, but there are caveats with this drug, including human safety and cost. Positive results have been observed usinng topical naltrexone in palette surgeonfish (*Paracanthurus hepatus*).
- Suggested approaches include improved water quality (with removal of activated carbon and protein skimming residual), improved diet with

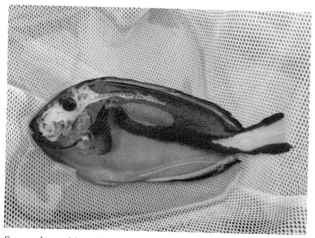

Figure 12.4　Severe lateral line depigmentation in a Pacific blue tang (*Paracanthurus hepatus*). RP Yanong

vitamin and mineral supplementation, sunlight exposure, and examination of habitat and social structure in addition to general diagnostic work up for pathogens.

OCULAR DISEASES

Ocular clinical signs and diseases are not uncommon but, in some cases, can be a challenge to accurately diagnose and manage. Imaging (radiography, ultrasound) may help rule in or out specific causes. Common signs and potential etiologies include:

- Exophthalmia
  - Supersaturation (resulting in gas pocket build up behind the eye; see **Figure 12.5**).
    - Correcting supersaturation in the system as well as careful removal of air pocket behind eye via needle/syringe may help, if system-related.

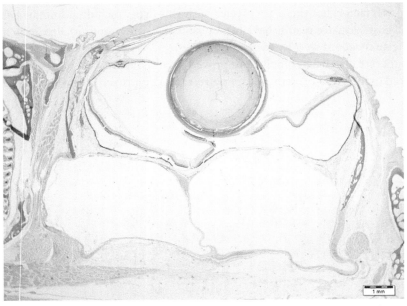

Figure 12.5   Histologic section demonstrating gas pockets (bottom of photo) located behind/below the eye of a milletseed butterflyfish (*Chaetodon miliaris*) with exophthalmia. No gas bubbles were seen in gill capillaries or on skin. RP Yanong

Figure 12.6 Exophthalmia in a tilapia (*Oreochromis* sp.) caused by *Streptococcus* sp. infection. RP Yanong

- Linear bubbles in gill vessels or on skin (other common clinical signs of supersaturation, or visible "excess bubbles" on the glass or other areas in the exhibit are not always present in cases of supersaturation).
- Infectious disease (see **Figure 12.6**).
- Mass/neoplasia.
- Acid/base disturbance.
- Corneal opacity.
  - Parasitic infection.
  - Water quality problem.
  - Trauma.
  - Nutrition.
- Hemorrhage.
  - Trauma.
  - Infectious disease (see **Figure 12.6**).

NUTRITIONAL DISEASES

- Body condition scoring (BCS) (described in Chapter 5; Clark et al., 2018) is a good tool for initial assessment. In some cases, problems may be as simple as underfeeding or overfeeding. However, other factors that may lead to non-ideal BCS include:

- Tank/pond competition.
- Bioavailability of nutrients (i.e., incorrect diet).
- Diet too protein or energy rich.
- Improperly stored or expired diet.
- Storage temperatures outside preferred ranges.
- Other water quality issues.
- Infectious or non-infectious diseases.
- Although specific pathologies may be indicative of nutritional disease, many of these may also be seen due to other etiologies, including toxins or infectious agents. Immunosuppression and stress resulting from poor nutrition may also facilitate infectious disease or lead to "poor doers" in general. **Tables 3.1** and **3.2** (Chapter 3) provide "general" disease signs associated with specific nutritional deficiencies, based on research in various fish species.
- Note (see Chapter 3, **Table 3.2**): cyprinids (including koi and goldfish) and other families that lack a true acid-secreting stomach may experience phosphorus deficiency (with resulting skeletal and opercular abnormalities) if fed a diet lacking inorganic phosphorus sources (e.g., if fed a diet with bone meal as the primary source of phosphorus and calcium; access to these minerals in bone meal requires the presence of acid).

CONSTIPATION

- Signs: Failure to defecate; anorexia.
- Cause: Using dry pelleted foods, especially in ponds.
- Treatment: add 0.3% magnesium sulfate salt to the diet.
- Laxatives: chopped earthworms, spinach, other vegetables.

BEHAVIORAL CONSIDERATIONS

- Know normals, including differences for juveniles and reproductively active males and females.
- Review water quality and nutrition, and rule-out infectious agents.
- Bottom-sitting (normal in benthic species, e.g., many catfishes, gobies, blennies, or sedentary/sleeping fish).
- Circling: damage to one eye or one fin; infection or damage to brain or inner ear; acute ammonia toxicity.
- Hovering (normal in angelfish, hatchetfish, Siamese fighting fish, some fancy goldfish).

- Occasional gulps of air at the surface: normal for anabantoid fishes (with the labyrinth organ); gouramis, bettas, paradisefish; *Corydoras* catfish (gulp air and use intestine as an accessory breathing organ); and some others.
- Upside down swimming (normal in upside down catfish).
- Be cognizant of potential for social or reproductive aggression often related to increased territoriality during and after spawning.

CHAPTER 13
HEALTH CERTIFICATES/CERTIFICATES OF VETERINARY INSPECTION

- If working with a retailer, wholesaler, and/or even an advanced hobbyist, the veterinarian may be asked to assist with health certificates for aquarium fish export/movement.
- US interstate transport of aquarium fish may not require a veterinary health certificate, but aquatic animal health regulations vary depending on species and state. Importing states should be contacted for verification.
- US veterinarians interested in assisting with aquatic animal export should receive Category II accreditation from the National Veterinary Accreditation Program and be familiar with the Veterinary Export Health Certification System (VEHCS). Please note: some countries do not allow electronic endorsement by the USDA APHIS VS and will require a hard copy, so more time will be required for the entire process. The USDA requires at least 48–72 hours to return documents.
- A Category II USDA accredited veterinarian may be asked to assist with issuing export health certificates for aquarium fish export/movement.
- International requirements for export health certificates for aquarium fish vary depending on animal origin, species, numbers, intended use, and destination. Some countries and some species may require specific pathogen testing and/or use of specific forms, whereas other countries are less stringent and may accept only visual examinations.
- Some countries only require the issuing veterinarian's signature, whereas others require both the issuing accredited veterinarian's signature as well as endorsement from the USDA APHIS VS.
- For questions regarding specific country import requirements, check the Import/Export International Regulations (IRegs) website (https://www.aphis.usda.gov/aphis/ourfocus/animalhealth/export/iregs-for-animal-exports/ct_iregs_animal_exports_home) and/or request an import permit from the importing country/final destination.

DOI: 10.1201/9781003057727-13

Health Certificates/Certificates of Veterinary Inspection

- For USDA endorsement services and locations, visit https://www.aphis.usda.gov/animal_health/contacts/field-operations-export-trade.pdf. Please note some locations require an appointment.
- Because live animal shipments may be held up if paperwork is not completed accurately, every effort should be given to make sure both the requirements and documentation are done correctly prior to animal movement.

(Thanks to Kathleen Hartman, USDA APHIS VS, for assistance with this section)

CHAPTER 14

DRUGS AND CHEMICALS

ADMINISTRATION METHODS AND FORMULARY

PRE-TREATMENT CONSIDERATIONS

- Before administering any drug in the water (immersion/bath route), discontinue chemical (e.g., carbon) filtration, protein skimming, and for aquarium facilities, ozone during treatment, as this will remove, alter, or inactivate the drug. The impact on the biofilter should also be considered and whether biofilter exposure to treatment can be avoided; however, logistics and target pathogen (e.g., parasites) may require inclusion of the biofilter, so water quality should be monitored. If possible, backwash and siphon detritus to reduce organic loading in the system. Adequate aeration is also important during any water treatment.
- Due to the aquatic nature, generally small size, and frequently large numbers of aquarium fish to be treated, a variety of atypical methods are utilized to deliver antibiotics to pet fish. Standard parenteral methods can and commonly are used to dose aquarium fish with antibiotics, but the clinician must also be familiar with the terminology applied to water borne treatments.

ROUTES OF DRUG AND CHEMICAL ADMINISTRATION FOR ORNAMENTAL FISH

- **Bath:** more appropriate for some disease issues (e.g., external parasites or primary external bacterial disease (e.g., columnaris)) but usually less desirable for internal/systemic infections than injectable or oral. Drug is dissolved in the water in which the fish are swimming. Exposure may vary from a 15-minute (short term) to 24-hour (long term) bath (drug and protocol dependent). Dosage is based on water volume, but water chemistry may also affect required dose (e.g., low hardness will facilitate uptake of oxytetracycline).
- Dosage is normally based on volume of water and not on fish biomass.*
  - Note: when antibiotics are used as bath treatments, ideally they should be used daily for at least 5–7 days (see **Table 14.1**). Water changes

Table 14.1 Antimicrobial Agents.

FISH FORMULARY – ANTI-MICROBIAL AND ANTI-FUNGAL (ANTI-OOMYCETE)

DRUG	ROUTE	DOSAGE	USE/NOTES
Acriflavine	Bath	4 mg/L x 4h* (Reimschuessel et al., 2021); 10 mg/L x 4h** (Plakas et al., 1998).	*Rainbow trout, **channel catfish, for water mold infections.
Acyclovir	ICe	10 mg/kg (Cardé et al., 2020).	For treating koi infected with cyprinid herpesvirus 3 (KHV). One treatment is safe, but multiple doses may be required.
Amikacin sulfate	IM; ICe	5 mg/kg IM q12h; 5 mg/kg IM q72h x 3 tx (Wildgoose and Lewbart, 2001); 5 mg/kg ICe q24h x 3d, then q48h x 2 tx (Johnson, 2006).	Renal efficiency should be evaluated if possible.
Amoxicillin	IM; IV; PO	12.5 mg/kg IM* (Brown and Grant, 1992); 25 mg/kg PO q12h (Stoskopf, 1999); 40 mg/kg IV q24h** (della Rocca et al., 2004a); 80 mg/kg PO q24h x 10 days** (della Rocca et al., 2004a); 40–80 mg/kg via feed x 10 days** (Noga, 2010); 110 mg/kg/day in feed*** (Ang et al., 2000).	*Atlantic salmon (Salmo salar), **sea bream (Sparus auratus), ***channel catfish (Ictalurus punctatus).
Ampicillin	IM; IV; PO	10 mg/kg IM q24h* (Treves-Brown, 2000); 10 mg/kg IV q24h** (Plakas et al., 1991); 50–80 mg/kg/day in feed x 10 days (Noga, 2010).	*Atlantic salmon (Salmo salar), **striped bass (Morone saxatilis).
Azithromycin	ICe; PO	30 mg/kg PO q24h x 14 days (Fairgrieve et al. 2005); 40 mg/kg ICe (Fairgrieve et al., 2006).	Chinook salmon (Oncorhynchus tshawytscha).
Aztreonam	ICe; IM	100 mg/kg ICe or IM q48h x 7 injections (Roberts et al., 2009).	For Aeromonas salmonicida in koi (Cyprinus carpio).

(Continued)

Table 14.1 (Continued)

FISH FORMULARY – ANTI-MICROBIAL AND ANTI-FUNGAL (ANTI-OOMYCETE)

DRUG	ROUTE	DOSAGE	USE/NOTES
Benzalkonium chloride	Bath	0.5 mg/L indefinitely (Treves-Brown, 2000); 10 mg/L x 30–60 minutes (Treves-Brown, 2000).	This compound is a quaternary ammonium that is broadly antimicrobial.
Bronopol	Bath	15-50 mg/L x 30–60 min bath (Pottinger et al., 1999; Wildgoose and Lewbart, 2001).	*Saprolegnia*/water mold infections, fish eggs.
Cefovecin	SC, IM	16 mg/kg SC (Seeley et al., 2016)*; 8 mg/kg IM (Steeil et al., 2014)**.	*Copper rockfish (*Sebastes caurinus*), **bamboo shark (*Chiloscyllium plagiosum*).
Ceftazidime	IM or SC	22 mg/kg IM, ICe q72–96h x 3–5 tx (Roberts, 2009).	Third generation cephalosporin that is generally effective against gram negative organisms.
Chloramine-T	Bath	2.5-20 mg/L x 60 min for up to 3d (Cross and Hursey, 1973; AADAP-FWS, 2022).	Disinfectant; used for bacterial gill disease and, in some cases, for external protozoans.
Diquat dibromide	Bath	2-18 mg/L for 1–4h x 1–4 tx q 24–48h; 19–28 mg/L for 30–60 min x 1–3 tx q48h (AADAP-FWS, 2022).	Controls columnaris disease in FW fishes.
Enrofloxacin	IM, ICe, PO, bath	5–10 mg/kg IM, ICe q48h x7–21d (dilute 1:1 with sterile saline to reduce irritation; Lewbart et al., 1997); 5–10 mg/kg PO 10–14d, or 0.1% in food 10–14d)Stoskopf, 1999); 2.5 mg/L x 5 hr bath q24h x5-7d, with 50%–75% water change between treatments (Lewbart et al., 1997); 10 mg/kg ICe x5d for koi at 20°C (Lewbart et al., 2005).	IM injections can lead to severe inflammation (Scott et al., 2020). For bath treatments, change at least 50% of water between doses. For a review of fluorinated quinolone use in fishes, see Samuelsen, 2006.

Table 14.1 (Continued)

FISH FORMULARY – ANTI-MICROBIAL AND ANTI-FUNGAL (ANTI-OOMYCETE)

DRUG	ROUTE	DOSAGE	USE/NOTES
Erythromycin (for PO, water soluble phosphate)	IM, ICe, PO	10–25 mg/kg IM, ICe; 75 mg/kg PO q24h x 7–10 days (FWS, 2022)*; 75 mg/kg PO q 24 h (di Salvo et al., 2014)**.	Gram-positive bacteria. Widely sold as an over-the-counter medication for aquarium fishes (but not therapeutic for the more common gram negative bacterial diseases).
Florfenicol	IM, PO	10 mg/kg IM q24h*, 10, 25, or 50 mg/kg PO q24h*△; 10 or 100 mg/kg IM q12h†; 10, 25, or 50 mg/kg PO q12h‡‡ (Yanong et al., 2005) 40–50 mg/kg PO, IM, ICe q12-24h˚ (Lewbart et al., 2005) 40 mg/kg IM (Zimmerman et al., 2006)**.	Koi: *MIC 1-6µg/mL; *△MIC 1, 3, 6 µg/mL; gourami: †MIC 1 or 6µg/mL; ‡‡MIC 1, 3, 6 µg/mL; ˚red pacu; **white-spotted bamboo shark.
Formalin	Bath	1 mL 100% formalin (37%–40% formaldehyde)/10 gallons (= 25 mg/L) 12–24 hr; repeat (Noga, 2010).	External *Saprolegnia*/water mold, which can be challenging to treat; NOT for bacterial infections;
Furazolidone	PO, Bath	PO: 1 mg/kg* (Plakas et al., 1994); 30 mg/kg** (Xu et al., 2006); 50–100 mg/kg q24h x 10–15 days (Noga, 2010)***. Bath: 1-10 mg/L for ≥24h (Noga, 2010).	Nitrofuran, bacterial infections; *channel catfish; **Nile tilapia; ***general.
Hydrogen peroxide 3% (30 mg/mL); 35% (350 mg/mL)	Bath	0.1 mL/L (appx 3.1 mg/L) x 1h (3%)* (Russo et al., 2007); 50mg/L x 1h**; 750–1000 mg/L x 15min xq24 until hatch*** (Phu et al., 2015).	*For external bacteria in swordtails; **for columnaris in channel catfish; ***for freshwater-reared warmwater finfish eggs.
Itraconazole	In feed	1–5 mg/kg q 24h in feed q1–7d (Stoskopf, 1999).	For systemic mycoses, but systemic mycoses are a challenge to manage.

(Continued)

Table 14.1 (Continued)

FISH FORMULARY – ANTI-MICROBIAL AND ANTI-FUNGAL (ANTI-OOMYCETE)

DRUG	ROUTE	DOSAGE	USE/NOTES
Kanamycin sulfate	PO, ICe, Bath	50 mg/kg q24h in feed; 20 mg/kg ICe q3d x 5 tx; 50-100 mg/L q72h x 3 treatments (Noga, 2010).	Toxic to some fish; for bath tx, do water change (50%–75%) between tx; absorbed from water.
Ketaconazole	PO, IM, ICe	2.5–10 mg/kg PO, IM, ICe (Stoskopf, 1999).	For systemic mycoses, but systemic mycoses are a challenge to manage.
Malachite green (Zn-free)	Bath	0.1 mg/L q3d x 3 tx (Noga, 2010).	Aquarium freshwater fish, for water mold infections; caution, mutagenic, teratogenic, toxic to some fish species and fry; increased toxicity at lower pH and higher temperatures; stains various materials.
Nitrofurazone	Bath	2–5 mg/L q24h x 5–10d; 50 mg/L for 3h (Wildgoose and Lewbart, 2001).	Nitrofuran; bacterial infections.
Oxolinic acid	PO, Bath	5–25 mg/kg PO q24hr* (Stoskopf, 1999); 25–50 mg/kg PO q24h** (Treves-Brown, 2000); 50mg/kg PO q24h x 5 days*** (Coyne et al., 2004a; Coyne et al., 2004b).	FW species*; SW species**; rainbow trout***.
Oxytetracycline	PO, IM, Bath	55–83 mg/kg PO q 24h x 10d; 7–10 mg/kg IM (red pacu; Doi et al., 1998); 60 mg/kg IM q7d (carp; Grondel et al., 1987); 400 mg/L (buffer if oxytetracycline hydrochloride; treat in reduced hardness if possible) (Vorback et al., 2019).	Antibiotic; oxytetracycline binds calcium, so higher total hardness reduced effectiveness in bath treatments.

Table 14.1 (Continued)

FISH FORMULARY – ANTI-MICROBIAL AND ANTI-FUNGAL (ANTI-OOMYCETE)

DRUG	ROUTE	DOSAGE	USE/NOTES
Nystatin-neomycin sulfate-thiostreptotriamcinolone acetodide (Panolog® ointment)	Topical	q12h for 30–60 sec with fish out of water, gills submerged (Lewbart, 2006).	Wounds.
Potassium permanganate	Bath	2 mg/L as an indefinite bath (Wildgoose and Lewbart, 2001).	Oxidizer; external bacteria; organics will deactivate – pinkish/purple color turns to brown when deactivated (may need to add an additional 2 mg/L if under treatment time); in earthen ponds, 4h tx; in tank systems, consider treatment times of 1–2 hrs but watch fish. Will burn away slime coat – consider following with a therapeutic salt bath of 2ppt.
Povidone iodine solution and ointment	Topical	1:10 dilution or 20–100 mg/L for 10 min; rinse immediately afterward. Thin layer of 10% ointment (Lewbart, 1999).	For various skin wounds including post-surgery.
Silver sulfadiazine cream	Topical	q12h for 30–60 sec with fish out of water, gills submerged (Lewbart, 1998b).	Wounds; dip fish in clean water prior to returning to life support system
Sulfadimethoxine/ ormetoprim	PO	50 mg/kg/day in feed x 5 days (Noga, 2010).	Top dress; not effective against columnaris disease.
Trimethoprim/ sulfamethoxazole	PO, Bath	30 mg/kg PO q24h x 10–14 days (Noga, 2010); 0.2% feed x 10–14 days (Noga, 2010); 20 mg/L x 5–12h bath q24h x 5–7 days (Bergiso and Bergiso, 1978).	A good broad-spectrum antibiotic; 50%–75% water changes between doses.
Triple antibiotic ointment	Topical	q12h for 30–60 sec with fish out of water, gills submerged (Lewbart, 1998b).	External bacteria/wounds; dip fish in clean water prior to returning to tank.

(70%–90%) should take place between treatments to improve water quality and remove drugs and metabolites for the next dosing. This protocol is much easier to follow in a home or hospital aquarium than in a pet store or wholesale facility, although facilities with good sources of treated municipal or well water may be able to do these relatively easily.
- **Dip:** refers to a treatment in which the fish is submerged in a particular solution for between 1 second and 15 minutes. Water volumes are usually smaller than those of bath treatments, and drug concentrations are frequently higher.
- **Flush or Flow Through:** requires constant water flow. Most frequently used in raceways or narrow vats. The chemotherapeutant is added to the inflow area at a given rate, and fish are exposed to the drug as it passes over them with the water current. This is similar to the dip procedure except that fish may not have to be removed from their normal holding area.
- **Indefinite Bath:** medication is added to tank/system and, usually, there is no water change or immediate retreatment.
- **Injection:** the antibiotic is given by injection with the aid of a hypodermic needle and syringe. Routes may be subcutaneous, intradermal, intramuscular, intravenous, and intracoelomic (in the body cavity, formerly "intraperitoneal"). Place fish in plastic bag, drain water, and inject through bag. Alternatively, fish can be sedated prior to injection.
  - **Subcutaneous/intradermal:** can be a challenge with small fish or fish with thin skin. For medium to large fish, position needle under a scale, through the scale pocket at an acute angle (e.g., 10–25 degrees, depth varying depending on the size of the fish).
  - **Intramuscular:** inject into the epaxial muscle a few mm on either side of the dorsal fin.
  - **Intravenous:** proceed as if collecting blood from the caudal vessels; once blood is seen entering the syringe, inject into the vessel(s).
  - **Intracoelomic (ICe) Injection:** inject either into the caudo-ventral abdomen or just caudal to the base of the pelvic fin a few mm from the bone (line). Be careful not to go too deep to avoid penetrating internal organs. Consider placing the fish in dorsal recumbency so that organs move away from the ventrum.
- **Oral:** medication is mixed with the food in order to treat the fish. This is usually done by incorporating the drug into a gelatinized food mixture or by mixing the drug into commercial feed using small amounts of fish or canola oil as a binder. For larger fish patients, medication can be placed into a chunk of food or prey item and then fed or force-fed to the patient.

- Manufactured gel food (can mix in medications): Mazuri® Aquatic Gel Diet.
- Recipe for gelatinized medicated food (modified from recipe of Dr. John Gratzek, College of Veterinary Medicine, University of Georgia):
  - Boil 500 ml tap water.
  - Add 21–35 g powdered unflavored gelatin (three to five 7-g packets), stir to dissolve, cool (do not set).

Mix in blender:

- 250 g flake food.
- 500 ml tap water.
- 25 ml cod liver oil + 25 ml vegetable oil (optional).
- Can of sardines, tuna, or baby food spinach (optional).
- Add medications and mix.
- Mix medicated food with gelatin, stir, refrigerate, or freeze.
- Use cheese grater to make bite-sized pieces.
- **Topical:** The medication is applied directly to the lesion or parasite.

FORMULARY

- No drugs are FDA-approved for ornamental pet fish.
- Use caution (pay attention to runoff onto plants, into other water sources/natural water bodies) when disposing of treated water or when changing water in outdoor ponds.
- Systemic bacterial infections are often secondary to stress, poor water quality, parasites/fungi/protists/viruses.
- Adding antibiotics to a tank can be deleterious to natural, beneficial bacteria (i.e., can upset the microbiome); it is preferable to use dips and, if treating in the water, to use a separate hospital tank. Ideally, treat based on culture and sensitivity results.
- Some drugs (e.g., copper) are highly toxic to corals and other reef invertebrates, algae, and plants. Salt at acceptable levels for fish (1–3 g/L) can be toxic to freshwater plants.
- The FDA is currently examining the availability of prescription drugs, especially antibiotics (for more information on the FDA and the use of drugs in aquaculture and fish, consult the following website: https://www.fda.gov/animal-veterinary/animal-health-literacy/aquaculture-and-aquaculture

-drugs-basics). Despite the passage of the Minor Use and Minor Species (MUMS) Animal Health Act (2004), we have not seen dramatic increases in the availability of approved or legal drugs for use in fish. Sound pharmacokinetics, efficacy, and safety studies to support clinical use of antimicrobials and other chemotherapeutants in the numerous aquarium fish species in the industry are still needed. Relatively few pharmacology studies have been reported for aquarium fishes. What little information exists is often based on clinical efficacy and *in vitro* trials using a number of different antimicrobials. An online database (https://www.fda.gov/animal-veterinary/tools-resources/phish-pharm) contains valuable information on pharmacokinetics in fish (FDA CVM) that may help provide some guidance.

DRUGS AND DOSAGES

(See **Tables 14.1–14.4**)
- The majority of the current information on drugs and chemicals used in aquarium fish has been extrapolated from the aquaculture literature. There are a number of reasons for this, most of which revolve around funding for sound pharmacokinetic research. In the US, there are currently only three antibiotics approved for use in fish intended for human consumption (Romet-30® – sulfadimethoxine/ormetoprim, Terramycin for Fish®– oxytetracycline, and, Aquaflor® – florfenicol). More studies targeting aquarium/non-food fish are needed. Much of the literature dealing with antibiotic usage in aquarium fish is empirical and anecdotal. Fortunately, the veterinarian treating aquarium fish can apply current extra label drug use regulations when selecting and initiating some, but not all, antibiotic therapy options.

The formulary is not meant to be a complete listing of all drugs available to treat and pharmacologically manage fishes nor has the information provided been proven to be safe and effective on all species. The formulary is a quick reference of drugs and dosages for the treatment and pharmacological management of pet fishes by a licensed veterinarian.

Table 14.2 Parasiticides

FISH FORMULARY – ANTI-PARASITIC

DRUG	ROUTE	DOSAGE	USE/NOTES
Acetic acid, glacial or vinegar	Bath	1–2 mL/L x 30–45 sec (glacial) (Wildgoose and Lewbart, 2001); Vinegar – increase dose based on solution concentration (e.g., if 4%, 25–50 mL/L x 30–45 sec).	Monogeneans, crustacean ectoparasites; safe for goldfish; may be toxic to smaller tropical fish.
Chloroquine diphosphate	Bath	10 mg/L, once, monitor for 21 days, repeat as needed (Lewis et al., 1988; Noga, 2010), remove with activated carbon if no relapse.	For *Amyloodinium*.
Copper sulfate (pentahydrate)	Bath	Freshwater fish: total alkalinity (TA, mg/L)/100 = mg/L copper sulfate (pentahydrate); e.g., if TA = 100, 100/100 = 1 -> 1 mg/L dose.	For freshwater, copper sulfate (pentahydrate) is 100% active; do not use if TA <50 mg/L; if TA>250 mg/L, use 2.5 mg/L dose or use chelated OTC products as recommended.
Copper $^{2+}$ (free copper; via copper sulfate (pentahydrate))	Bath	Saltwater fish: 0.15–0.20 mg/L free copper (Cu^{2+}) x14–21d for *Amyloodinium* (Noga, 2010); x 4–6 weeks for *Cryptocaryon* (caution: toxic to elasmobranchs and invertebrates).	External protists; for SW fish, target dose is based on free copper (Cu^{2+}) and should be reached gradually, over 2–4 days, to allow fish to upregulate copper detoxification mechanisms in the liver; monitor levels daily and add more as necessary to maintain 0.15–0.20 mg/L.
Diflubenzuron (Dimilin®)	Bath	0.01 mg/L q48h x 7d x 3 treatments, 7d apart (very effective but may kill desirable invertebrates). May requires an EPA pesticide applicator license to administer (Stoskopf, 1993). Anecdotal reports lufenuron (Program®) at a similar dose with good success.	Parasitic crustaceans (fish lice, anchorworm).
Dimethyl phosphonate; trichlorfon (liquid form for cattle grubs)	Bath	0.25–0.5 mg/L q10d x 3, with 20%–30% water change 24–48 hr post-treatment Caution: toxic to some species, including Serrasalmidae (pacu, piranha, redhooks, silver dollars).	Parasitic crustaceans (fish lice, anchorworm), leeches, monogeneans.
Dimetridazole	Bath	80 mg/L for 24h x 3 - 5 treatments (Yanong et al., 2004 and unpublished data).	Some effectiveness for reducing *Cryptobia iubilans*, but *C. iubilans* can be very difficult to treat
Emamectin benzoate	PO	5 µg/kg PO q24h* x 7 days (Hanson et al., 2011); 50 µg/kg PO q24h** x 7 days (Hanson et al., 2011).	Koi/*Argulus**; goldfish/argulus**; Atlantic salmon/sea lice spp.**
Fenbendazole	PO, Bath	PO: 200–250 mg/100 g food x3d, repeat in 2 wks (Whitaker et al., 1999); 50 mg/kg PO, repeat in 2 wks (Whitaker et al., 1999); Bath: 2 mg/L q7d x 3 tx (Noga, 2010).	Intestinal nematodes.

(Continued)

Table 14.2 (Continued)

FISH FORMULARY – ANTI-PARASITIC

DRUG	ROUTE	DOSAGE	USE/NOTES
Freshwater	Dip	Saltwater species: freshwater dip (pH adjusted to 8.0–8.5 if below 8.0) for 4–5 min. Aerate well and monitor very closely. Certain smaller fishes may not survive this treatment. If possible, test treatment on one fish first. Brackish water species: for marine parasites, some brackish water species (e.g., *Scatophagus* spp., *Monodactylus* spp.) can survive for extended periods of time in waters of higher total hardness.	External protists and monogeneans.
Formaldehyde 37% (100% formalin)	Bath or dip	Bath: 20–25 mg/L (ppm) (1.0 ml of 100% formalin [37% formaldehyde] in 10 gal [38 L] water) x 12–24 hr, then 50% water change, every other day x 3 treatments. Encysted parasites like "ich" and *Cryptocaryon* require several treatments (Noga, 2010). Or dip 100–250 ppm 30–60 min. Effective for some ecto-parasites in koi as a 10-min dip at 100–150 mg/L. Caution: carcinogen; may compromise biological filter; removes O_2 from water (increase aeration). Formalin is the only parasiticide approved for use in food fish. Monitor fish closely for signs of distress. Always change water between treatments. Caution: carcinogen – wear gloves and measure in well-vented area.	Protists; monogeneans; saprolegnia *Saprolegnia* (oomycete/water mold); formalin is approved for use in food fish; monitor fish closely for signs of distress; always change water between treatments; caution: carcinogen – wear gloves and measure in well-vented area).
Freshwater	Dip	Saltwater species: freshwater dip (pH adjusted to 8.0–8.5 if below 8.0) for 4–5 min. (Noga, 2010). Aerate well and monitor very closely. Certain smaller fishes may not survive this treatment. If possible, test treatment on one fish first (biotest). Brackish water species: for marine parasites, some brackish water species (e.g., *Scatophagus* spp., *Monodactylus* spp.) can survive for extended periods of time in waters of higher total hardness.	External protists and monogeneans.
Glacial acetic acid	Dip	2 ml/L 30–45 sec (test one fish for sensitivity).	Monogeneans on goldfish; may be effective against parasitic crustaceans (fish lice, anchorworm) (see acetic acid).
Hydrogen peroxide (3%; 30 mg/mL)	Dip or bath	0.22 mL/L (= 6.6 mg/L) x 1–2h*; 2.5–3.33 mL/L (= 75–100 mg/L) x 30min** q6d x 2tx; 10 mL/L (= 300 mg/L) x 10min*** (Russo et al., 2007).	Protists, monogeneans; species sensitivities; not recommended for use in gouramis and suckermouth catfish; *external flagellates in swordtails; **Pacific threadfin (marine) against *Amyloodinium*; ***kingfish (marine) against monogeneans.
Levamisole phosphate	Bath	1–2 mg/L x 24h*, 50 mg/L x 2h* (Harms, 1996) 4 g/kg feed q7d x 3 tx*/** (Harms, 1996).	Internal nematodes*; external monogeneans**.
Lufenuron	Bath	0.13 mg/L prn (Roberts et al., 2009)	Crustacean ectoparasites.

(Continued)

Table 14.2 (Continued)

FISH FORMULARY – ANTI-PARASITIC

DRUG	ROUTE	DOSAGE	USE/NOTES
Malachite green (zinc-free)	Bath	0.1–0.15 ppm for 12–24 hr +/- 20 ppm formaldehyde (Noga, 2010); 30% water change, then repeat. Repeat as long as parasites are present. Caution: carcinogen and teratogen – wear gloves. Toxic to scaleless fish.	"Ich;" fungal disease Caution: carcinogen and teratogen – wear gloves. Toxic to scaleless fish.
Metronidazole	Bath, PO	Bath: 6.6 mg/L q24h x3d, 25%–50% water changes between treatments (Noga, 2010).; 50 mg/kg body weight or 10 mg/g feed (top dressed) q24 x 5d (Whaley and Francis-Floyd, 1991). Bioencapsulation in *Artemia* (brine shrimp). Add 5 g metronidazole plus 15 mL brine shrimp in 500 mL saltwater for 0.25 hr will yield 9.32 μg metronidazole per shrimp (2500 μg per g of shrimp). 15 mL of strained live adult brine shrimp is approximately 16 g wet weight (≈ 262 shrimp/g) (Allender et al., 2011).	External/internal flagellates; gastrointestinal Spironucleus infestations can be treated via bath, but feed is preferred.
Potassium permanganate	Bath	1.3 mg/L 2x over 3 days* (Steckler and Yanong 2013); 2 mg/L as an indefinite bath** (Yanong unpublished data); 5 mg/L x 30-60min; 10 mg/L x 30 min*; 100 mg/L x 5–10min; 1 g/L x 10–40sec.	Freshwater fish only. See Table 14.1 comments; *may be effective against adult and juvenile stages of some *Argulus* spp.; ** some external protists (e.g., *Trichodina, Ichthyobodo*); can be more toxic at higher pHs; do not mix with formalin; species sensitivities – can be toxic to goldfish and other tropicals; biotest.
Praziquantel tablets and injectable solution	ICe, PO, Bath	5 mg/kg fish ICe or PO in food, q7–21d x 2–3tx (Lewbart, 1998a); 5–10 mg/L bath x 3–6 hr, repeat in 7d; Bioencapsulation in *Artemia* (brine shrimp). Add 15 mL of strained live adult brine shrimp and 2.5 g praziquantel per 500 mL volume for 30 min to achieve 8.6 μg praziquantel per shrimp (Allender et al., 2012).	Monogeneans; internal cestodes or internal digenean trematodes; note: for encysted worms, may incite severe inflammatory response; not very soluble in water – consider dissolving in ethanol first.
Salt (NaCl), solar salt, seawater, or artificial sea salt; avoid salts with anti-caking agents or other additives (i.e., Sodium ferrocyanide)	Dip or bath	Freshwater species: 30–35 g/L x 4–5 min*(Lewbart, 1998a); 1 g/L**; 2–3 g/L ***; 4–5 g/L ****; 4.8 g/L seawater***** (Noga, 2010).	* Tolerated by many species. ** For general osmoregulatory support; *** for some protists (*Chilodonella*); ****for tolerant species for *Ichthyophthirius* (for 1.5–2 life cycle duration); ***** for 30 days may be effective against *Lernaea* (anchorworm).
Trichlorfon (Neguvon® and Malthion®)	Bath	0.5 mg/L, treatment is usually repeated several times with 25%–30% water changes between treatments (Lewbart, 1998b). A biotest is strongly recommended when using these compounds.	Effective in controlling crustacean ectoparasites.

CHAPTER 14: DRUGS AND CHEMICALS

Table 14.3

FISH FORMULARY – SEDATION, ANALGESIA, ANESTHESIA, AND EUTHANASIA*

DRUG	ROUTE	ANALGESIA/SEDATION/ ANESTHESIA	EUTHANASIA* (SEE NOTE BELOW TABLE)	USE/NOTES
Alfaxalone	Bath	0.5 mg/L for sedation*; 5 mg/L for anesthesia* (Bauquier et al., 2013; Minter et al., 2014; Bugman et al., 2016)**.	–	*Goldfish (Carassius auratus); **lowercase oscar (Astronotus ocellatus).
Atipamezole	IM	0.2 mg/kg IM (Fleming et al., 2003).	–	Dosing volume may be dependent on dexmedtomidine injection (Gaskins and Boylan, 2019)
Benzocaine	Bath	15–40 mg/L for transport sedation (Noga, 2010) 50–500 mg/L for anesthesia (Noga, 2010) 100–200 mg/L* for anesthesia (Skar et al., 2017).	–	*Lumpfish (Cyclopterus lumpus) less than 400 g.
Butorphanol	IM	0.4 mg/kg IM* (Harms et al., 2005); 10 mg/kg IM** *Baker et al., 2013). Analgesia.	–	*Koi (Cyprinus rubrofuscus) post-operative analgesia; ** koi (Cyprinus carpio) post-operative analgesia. Patient may experience decreased opercular rate. Lower dosing suggested.
Carbon dioxide	Bath	Not recommended.	Immerse in CO_2 saturated water; CO_2 will drop pH significantly; some fish may become hyperactive prior to unconsciousness (AVMA, 2020).	CO_2 from a source that can be regulated (i.e., CO_2 cylinders) is AVMA acceptable; use in well-ventilated area.

Table 14.3 (Continued)

FISH FORMULARY – SEDATION, ANESTHESIA, AND EUTHANASIA*

DRUG	ROUTE	ANALGESIA/SEDATION/ ANESTHESIA	EUTHANASIA* (SEE NOTE BELOW TABLE)	USE/NOTES
Clove oil (1:10 with 95% ethanol; stock approx. 100 mg/mL)	Bath	25–120 mg/L are effective in freshwater and marine species and results are comparable to MS-222, except that recovery may be prolonged (Pattanasiri et al., 2017); 500 mg/L* (Park, 2019).	10x upper anesthesia range for euthanasia.	Clove oil is a non-uniform mixture of isoeugenol, eugenol, methyleugenol, and others. To make a stock solution of 100 mg/mL, combine 1 part clove oil with 9 parts 95% ethanol (eugenol is poorly soluble in water); recovery may be prolonged; start with the lower end of this range; many bony fishes are anesthetized with 25–50 mg/L *for anesthesia of amur catfish (*Silurus asotus*), resulting in rapid (<1 min) induction time.
Dexmedetomidine	IM	0.025–0.1 mg/kg (Gaskins and Boylan, 2019).	—	For analgesia in elasmobranchs
Dexmedetomidine/ Ketamine/ Midazolam	IM	0.05–1.0 mg/kg dexmedetomidine + 2–4 mg/kg ketamine + 0.2 mg/kg midazolam IM (Christiansen et al., 2014).	—	Sedation in black sea bass (*Centropistis striata*). Use with caution. Fatal to red porgy (*Pagrus pagrus*).
Ethanol	Bath	Not recommended.	10–30 mL of 95% ethanol/L (AVMA, 2020).	Immerse to effect.

(Continued)

Table 14.3 (Continued)

FISH FORMULARY – SEDATION, ANESTHESIA, AND EUTHANASIA*

DRUG	ROUTE	ANALGESIA/SEDATION/ ANESTHESIA	EUTHANASIA* (SEE NOTE BELOW TABLE)	USE/NOTES
Etomidate	Bath	0.5–3.6 mg/L (Plumb et al., 1983). Anesthesia.	–	Good for cyprinids and channel catfish (*Ictalurus punctatus*). Lower doses for striped bass (*Morone saxatilis*).
Eugenol (Aqui-S20E is 10% eugenol)	Bath	10–100 mg/L eugenol (AADAP, 2021).	10x upper anesthesia range for euthanasia.	Eugenol is one component of clove oil.
Hydromorphone	IM	0.2 mg/kg IM (Gaskins and Boylan, 2019).	–	Post-operative analgesia.
Isoflurane	Bath/Vaporize	0.5–2.0 mL/L to effect (Harms, 1999). Anesthesia.	–	Can be difficult to control, and there is a risk to personnel.
Ketamine	IM	5–10 mg/kg IM* (Gaskins and Boylan, 2019); 66–88 mg/kg** (Stoskopf, 1999). Anesthesia	66–88 mg/kg followed by pentobarbital	* Usually combined with other drugs such as dexmedetomidine and midazolam; **immobilization for short procedures, recovery may be >1h
Ketoprofen	IM	2 mg/kg IM* (Harms et al., 2005); 8 mg/kg IM** (Greene et al., 2020).	–	*Koi (*Cyprinus carpio*) post-operative analgesia. **Nile tilapia (*Oreochromis niloticus*) and rainbow trout (*Onchorhynchus mykiss*) for analgesia.
Lidocaine	IM	1–2 mg/kg (Harms, 1999).	–	For topical analgesia. Efficacy and safety not widely tested.

Table 14.3 (Continued)

FISH FORMULARY – SEDATION, ANESTHESIA, AND EUTHANASIA*

DRUG	ROUTE	ANALGESIA/SEDATION/ ANESTHESIA	EUTHANASIA* (SEE NOTE BELOW TABLE)	USE/NOTES
Metomidate hydrochloride	Bath, PO	Sedation: 0.06–0.2 mg/L* (Stoskopf, 1995); 1 mg/L** (Kilgore et al., 2009); 0.5–1 mg/L and 2.5–5 mg/L*** (Harms, 1999); Anesthesia: 5–10 mg/L‡ (Harms, 1999) 7 mg/kg PO (turbot, *Scophthalmus maximus*); Hansen et al., 2003.	Must be buffered; 40–100 mg/L; small (5.1–7.6 cm TL) goldfish, koi, and some cichlids were successfully euthanized in this one-step procedure; some catfishes require higher doses (*Otocinclus* – 250 mg/L; *Corydoras aeneus* – 1000 mg/L) (Yanong, 2021)	Gouramis may be sensitive; do not use with cichlids if pH<5. Sedation: *transport sedation; **convict cichlid (*Amatitlania nigrofasciata*) 24h transport sedation; ***light and heavy sedation; fish may still demonstrate reactivity. Anesthesia‡: monitor respiration carefully and switch to or add unmedicated water if too deep to avoid overdose; not recommended for long procedures; fish may demonstrate reactivity.
Morphine	IM	5 mg/kg (Baker et al., 2013). Analgesia.	Not recommended as one step.	Koi (*Cyprinus rubrofuscus*)
MS-222	–	–	–	See tricaine methanesulfonate.
Pentobarbital	IV, IC, or ICe	Not recommended.	60-100 mg/kg.	–
2-Phenoxyethanol	Bath	0.1-0.6 mL/L* (Treves-Brown, 2000); 0.15 mL/L** (Penning et al., 2017).	≥ 0.6 mL/L (or ≥ 0.4 mg/L).	*Carp; **elasmobranchs.

(Continued)

Table 14.3 (Continued)

FISH FORMULARY – SEDATION, ANESTHESIA, AND EUTHANASIA*

DRUG	ROUTE	ANALGESIA/SEDATION/ ANESTHESIA	EUTHANASIA* (SEE NOTE BELOW TABLE)	USE/NOTES
Potassium chloride	Intracardiac	–	10 mmol/kg (750 mg/kg) intracardiac (Louis et al., 2020).	Secondary method following immersion or injectable anesthesia.
Propofol	IV, Bath	3.5-7.5 mg/kg IV* (Fleming et al., 2003). Anesthesia 7 mg/L bath** (GholipourKanani and Ahadizadeh, 2013). Anestheia 2.5-5 mg/L*** (Oda et al., 2014, 2018). Anesthesia	Not recommended as 1 step	*Gulf of Mexico sturgeon (*Acipenser oxyrinchus de soti*); **goldfish; ***koi (but for induction/ short procedures only) (Oda et al., 2014).
Quinaldine sulfate	Bath	25-100 mg/L (Harms, 1999; Treves-Brown, 2000). Anesthesia.	Not recommended as one step.	Many species, biotest suggested.
Robenacoxib	IM	2 mg/kg (Raulic et al., 2020). Analgesia.		Rainbow trout (*Oncharhynchus mykiss*).
Sevoflurane	Bath	–	5–20 mL/L (AVMA, 2020).	Risk to personnel necessitates proper ventilation.
Sodium bicarbonate	Bath	Not recommended.	30g/L (Noga, 2010).	CO_2 produced; only use when no other methods/agents available; not an approved AVMA method.

Table 14.3 (Continued)

FISH FORMULARY – SEDATION, ANESTHESIA, AND EUTHANASIA*

DRUG	ROUTE	ANALGESIA/SEDATION/ ANESTHESIA	EUTHANASIA* (SEE NOTE BELOW TABLE)	USE/NOTES
Sodium bicarbonate tablets (Alka-Seltzer, Bayer)	Bath	Not recommended.	2–4 tablets/L (Gratzek et al., 1992).	CO_2 produced; only use when no other methods/agents available; not an approved AVMA method.
Tricaine methanesulfonate, MS-222 (10 mg/ml stock solution) buffered prior to use	Bath	15–50 mg/L* (Harms, 1999); 50–200 mg/L, induction (Harms, 1999); 50–100 mg/L maintenance**; 1 g/L spray*** (Noga, 2010).	250–500 mg/L, or 10x upper anesthetic dose**** (AVMA, 2020).	Must be buffered with sodium bicarbonate: MS-222 (1:1 or 1:2) or sodium carbonate: MS-222 (0.5:1 or 1:1). Unbuffered MS-222 may be stressful to fish and lead to a missed protozoal parasite diagnosis (Callahan and Noga, 2002); *sedation; **anesthesia; ***large fish; anesthesia; spray onto gills with aerosol pump sprayer; ****goldfish, koi, cichlids, and some other species may require a second step/adjunctive method after fish is unconscious (e.g., pithing) (Balko et al., 2018).

* Most agents listed are included in the AVMA Guidelines on Euthanasia of Animals, 2020. S6. Fish and Aquatic Invertebrates. Refer to these guidelines for more specific information on approved methods. For immersion methods, it is now recommended to keep fish in euthanasia bath for a minimum of 30 minutes after loss of opercular activity/respiration and reactivity to stimuli.

Table 14.4

FISH FORMULARY – MISCELLANEOUS

DRUG	ROUTE	DOSAGE	USE/NOTES
Ascorbic acid (vitamin C)	IM or SC	3–5 mg/kg IM 2 24h (Saint-Erne, 1992); 25–50 mg/kg IM, SC (Gaskins and Boylan, 2019).	Dilution recommended prior to injection to reduce irritation.
Atropine	IM, IV, ICe	0.1 mg/kg ICe, IM, IV (Stoskopf, 1993).	For organophosphate or chlorinated hydrocarbon exposure.
B-vitamin complex	SC	10 mg/kg SC (Gaskins and Boylan, 2019).	May serve as an appetite stimulant.
Bacaplermin	Topical	Place a thin layer on lateral line depigmentation area for 3 min (Fleming et al., 2008).	Successful treatment may require a change in the environmental parameters.
Beta-glucans	PO	1 g/kg of feed × 24 days fed at 3% BW (Russo et al., 2006)*; 2 g/kg feed for 7 days (Siwicki et al., 1994)**.	*Immunostimulant. Effective in the cyprinid red-tailed black shark (*Epalzeorhynchos bicolor*) when challenged with *Streptococcus iniae*. It is recommended to only use beta-glucans and other immunostimulants temporarily before/around periods of stress and not on a long-term basis. **To increase immunocompetency in rainbow trout (*Onchorhynchus mykiss*).
Carbon (activated)	Water treatment	2 g/L of aquarium water (Noga, 2010).	Removed medications and organic compounds from the life support system. Discard carbon after 14 days.
Carp pituitary extract	IM	0.75–3.0 mg/kg IM (Treves-Brown, 2000); 5 mg/kg IM combined with human chorionic gonadotropin (Stoskopf, 1999).	Dosing depends on size and sex of fish. Do not administer unless the ova are mature.

Table 14.4 (Continued)

FISH FORMULARY – MISCELLANEOUS

DRUG	ROUTE	DOSAGE	USE/NOTES
Chlorine/chloramine neutralizing agents	In water	Use as directed. Also see sodium thiosulfate.	Removes and/or binds chlorine and/or chloramine from water.
Dexamethasone	ICe, IV or IM	ICe (Lewbart, 1998b).	Chlorine/chloramine toxicity; shock.
Doxapram	ICe, IV, topical	5 mg/kg ICe, IV (Stoskopf, 1993); 5 mg/kg topical on gills (Gaskins and Boylan, 2019).	Respiratory stimulant of questionable value.
Epinephrine (1:1000)	ICe, IM, IV, intracardiac	0.2–0.5 ml ICe, IM, IV, intracardiac (Stoskopf, 1993).	Shock/cardiac arrest.
Furosemide	ICe, IM	2–5 mg/kg ICe, IM (Stoskopf, 1999).	For treating ascites. Of questionable value due to fish lacking a Loop of Henle.
sGnRHa (salmon gonadotropin-releasing hormone analogue) + domperidone (Ovaprim®, Syndel USA)	ICe, IM	0.5 mL/kg (0.5 µL/g) ICe, IM; for some species/situations, a 10% primer dose is provided followed by a 90% resolving dose at least 6 hrs later (Hill et al., 2009; Yanong et al., 2009).	To induce/assist with spawning/release of mature gametes (mature eggs/sperm); generally occurs 4–12 hours post-injection for warm-water species; see product label.
Haloperidol	IM	0.5 mg/kg IM (Stoskopf, 1999).	A dopamine blocker that stimulates ovulation. Use with luteinizing releasing hormone analog (LRH-A).
Hetastarch	IV	0.5–1.0 mL/kg slowly IV (Gaskins and Boylan, 2019).	Colloidal replacement.

(Continued)

Table 14.4 (Continued)

FISH FORMULARY – MISCELLANEOUS

DRUG	ROUTE	DOSAGE	USE/NOTES
Human chorionic gonadotropin (hCG)	IM	20 U/kg IM, repeat in 6 hr* (Stoskopf, 1999); 23–232 U/kg IM for males and 30–828 IU/kg IM for females, repeat every 6 hr for up to three treatments** (Chorulon; Stoskopf, 1999).	*Combined with carp pituitary extract. **Note the wide range of concentrations. Doses vary by species and sex. Consult appropriate experts or literature and product label.
Hydrocortisone	ICe, IM	1–4 mg/kg ICe, IM (Stoskopf, 1999).	For chronic stress or shock.
Hydrogen peroxide (3%)	Bath	0.25 mL/L (Noga, 2010)	For acute hypoxia.
Luteinizing releasing hormone analog (LRH-A)	IM	2 µg/kg IM initially followed by 8 µg/kg 6 hr later (Stoskopf, 1999).	Can be given with haloperidol or reserpine to stimulate ovulation.
Methyltestosterone	PO	30–60 mg/kg PO (Guerrero, 1975; Johnstone et al., 1983; McGeachin et al., 1987; Varadaraj and Pandian, 1989; Reimschuessel et al., 2022).	For masculinization of females for aquaculture purposes. Dosing and frequency vary by species (usually tilapia and rainbow trout).
Naltrexone	Topical	Combine 4 mg naltrexone in 10g iLEX petroleum paste and apply a 1–3 mm thick layer to affected area (Strobel et al., 2023).	For lateral line depigmentation (LLD) in surgeonfish (*Paracanthurus hepatus*).
Nitrifying bacteria	Immersion	Use as directed (commercial products). Material (biomedia etc.) from an existing filter can be used to "seed" a naïve filter.	To "seed" a biological filter. Many commercial products available. Efficacy varies. Biosecurity risk if donor system has infectious disease problems.

FISH FORMULARY – MISCELLANEOUS

DRUG	ROUTE	DOSAGE	USE/NOTES
Nucleotide	PO	2 g/kg of feed × 24 days fed at 3% BW (Russo et al., 2006).	Immunostimulant. Effective in the cyprinid red-tailed black shark (*Epalzeorhynchos bicolor*) when challenged with *Streptococcus iniae*. It is recommended to only use beta-glucans and other immunostimulants temporarily before/around periods of stress, and not on a long-term basis.
Oxygen	Topical/bath	Fill and seal plastic bag containing fish with two-thirds 100% oxygen and one-third clean water (Lewbart, 1998b).	For acute environmental hypoxia; close bag tightly with rubber band or zip tie; keep fish in bag until normal behavior returns.
Reserpine	IM	50 mg/kg IM (Stoskopf, 1999).	Use with LRH-A to stimulate ovulation.
Sodium chloride (salt)	Bath	1-3 g/L bath (Lewbart, 1995); 3-5 g/L (Noga, 2010).	For freshwater fish to reduce stress-induced mortality; artificial sea salts preferred; use non-iodized table salts; some anticaking agents in solar salts are highly toxic; highly variable species sensitivity to salt (some catfish and mormyrids (e.g., elephantnose fish) may be sensitive); toxic to some plants.
Sodium thiosulfate	Bath	Use as directed to neutralize chlorine and/or chloramine; 10 g/1000L (removes 2 mg/L chlorine); Lewbart, 1998b; 100mg/L for acute chlorine exposure (Stoskopf, 1993).	This is the active ingredient in many chlorine/chloramine neutralizers; ammonia released by detoxification of chloramine is removed by functioning biological filter or chemical means (see zeolite).
Zeolite	Immersion	Use as directed to remove toxins; 20 g/L LSS water (Noga, 2010).	An ion-exchange resin that replaces sodium ions with ammonia; effective for removal of compounds such as antimicrobials (Ötker and Akmehmet-Balcıoğlu, 2005; Braschi et al., 2010; Homem and Santos, 2011).

BIBLIOGRAPHY

AADAP-FWS FDA INAD. Web site: https://www.fws.gov/fisheries/aadap/inads-available/index.html. Accessed November 21, 2022.

AADAP/FDA. https://www.fws.gov/fisheries/aadap/inads-available/immersion/diquat/. Accessed November 12, 2022.

AADAP FWS. https://www.fws.gov/inad/chloramine-t-9321. Accessed November 12, 2022.

Adamovicz L, Trosclair M, Lewbart GA. 2017. Biochemistry panel reference intervals for juvenile goldfish (*Carassius auratus*). *J Zoo Wild Med* 48(3):776–785.

Allender MC, Kastura M, George R, et al. 2010. Bioencapsulation of praziquantel in adult *Artemia*. *J Bioanal Biomed* 2:96–99.

Allender MC, Kastura M, George R, et al. 2011. Bioencapsulation of metronidazole in adult brine shrimp (*Artemia* sp). *J Zoo Wildl Med* 42(2):241–246.

Allender MC, Kastura M, George R, et al. 2012. Bioencapsulation of fenbendazole in adult *Artemia*. *J Exot Pet Med* 21(3):207–212.

Ang CY, Liu FF, Lay JO Jr, et al. 2000. Liquid chromatographic analysis of incurred amoxicillin residues in catfish muscle following oral administration of the drug. *J Agric Food Chem* 48(5):1673–1677.

Aquacalm® (metomidate) product label.

Aquaflor label. see http://www.fda.gov/AnimalVeterinary/DevelopmentApprovalProcess/Aquaculture/ucm132954.htm.

Arnold J. 2005. Hematology of the sandbar shark (*Carcharhinus plumbeus*). *Vet Clin Pathol* 34(2):115–123.

American Veterinary Medical Association. AVMA guidelines for the euthanasia of animals. 2020 edition. https://www.avma.org/sites/default/files/2020-01/2020-Euthanasia-Final-1-17-20.pdf. Accessed November 21, 2022.

Bailey KM, Minter LJ, Lewbart GA, et al. 2014. Alfaxalone as an intramuscular injectable anesthetic in koi carp (*Cyprinus carpio*). *J Zoo Wildl Med* 45(4):852–858.

Balko JA, Oda A, Posner LP. 2018. Use of tricaine methanesulfonate or propofol for immersion euthanasia of goldfish (*Carassius auratus*). *J Am Vet Med Assoc* 252(12):1555–1561.

Baker TR, Baker BB, Johnson SM, Sladky KK. 2013. Comparative analgesic efficacy of morphine sulfate and butorphanol tartrate in koi (*Cyprinus carpio*) undergoing unilateral gonadectomy. *J Am Vet Med Assoc* 243(6):882–890.

Bauquier SH, Greenwood J, Whittem T. 2013. Evaluation of the sedative and anaesthetic effects of five different concentrations of alfaxalone in goldfish, Carassius auratus. *Aquaculture* 396–399: 119–123.

Bergjso T, Bergjso HT. 1978. Absorption from water as an alternative method for the administration of sulfonamides to rainbow trout, *Salmo gairdneri*. *Acta Vet Scand* 19:102–109.

Braschi I, Blasioli S, Gigli L, et al. 2010. Removal of sulfonamide antibiotics from water: Evidence of adsorption into an organophilic zeolite Y by its structural modifications. *J Hazard Mater* 178(1–3):218–225.

Brito FMM, Claudiano G, Yunis J, et al. 2015. Hematology, biochemical profile and thyroid hormones of four species of freshwater stingrays of the genus *Potamotrygon*. *Braz J Vet Res Anim Sci, São Paulo* 52(3):249–256.

Brown AG, Grant AN. 1992. Use of ampicillin by injection in Atlantic salmon broodstock. *Vet Rec* 131(11):237.

Buchmann K, Bjerregaard J. 1990. Mebendazole treatment of psuedodactylogyrosis in an intensive eel-culture system. *Aquaculture* 86(2–3):139–153.

Bugman AM, Langer PT, Hadzima E, et al. 2016. Evaluation of the anesthetic efficacy of alfaxalone in oscar fish (*Astronotus ocellatus*). *Am J Vet Res* 77(3):239–244.

Cain DK, Harms CA, Segars A. 2004. Plasma biochemistry reference values of wild-caught southern stingrays (*Dasyatis americana*). *J Zoo Wildl Med* 35(4):471–476.

Callahan HA, Noga EJ. 2002. Tricaine dramatically reduces the ability to diagnose protozoan ectoparasite (*Ichthyobodo necator*) infections. *J Fish Dis* 25(7):433–437.

Cardé EMQ, Yazdi Z, Yun S et al. 2020. Pharmacokinetic and efficacy study of acylclovir against Cyprinid herpesvirus 3 in *Cyprinus carpio*. *Front Vet Sci* 7:587952.

Chen CY, Getchel RG, Wooster GA, et al. 2004. Oxytetracycline residues in four species of fish after 10-day oral dosing in feed. *J Aquat Anim Health* 16(4):208–219.

Chorulon (Human Chorionic Gonadotropin; HCG) Product Label. Merck Animal Health, Summit, NJ.

Christiansen E, Mitchell JM, Harms CA, et al. 2014. Sedation of red porgy (Pagrus pagrus) and black sea bass (Centropistis striata) using ketamine, dexmedetomidine and midazolam delivered via intramuscular injection. *J Zoo Aquar Res:* 2(3)62–68.

Collins S, Dornburg A, Flores JM, Dombrowski DS, et al. 2016. A comparison of blood gases, biochemistry, and hematology to ecomorphology in a health assessment of pinfish (*Lagodon rhomboides*). *PeerJ* 4:e2262 https://doi.org/10.7717/peerj.2262.

Coyne R, Samuelsen O, Kongshaug H, et al. 2004a. A comparison of oxolinic acid concentrations in farmed and laboratory held rainbow trout (*Oncorhynchus mykiss*) following oral therapy. *Aquaculture* 239(1–4):1–13.

Coyne R, Bergh O, Smith P, et al. 2004b. A question of temperature related differences in plasma oxolinic acid concentrations achieved in rainbow trout (*Oncorhynchus mykiss*) under laboratory conditions following multiple oral dosing. *Aquaculture* 245(1–4):13–17.

Cross DG, Hursey PA. 1973. Chloramine-T for the control of *Icthyophthirius multifiliis* (Fouquet). *J Fish Dis* 10:789–798.

Crosby TC, Kittel EC, Gieseker CM. 2022. Phish-pharm: A searchable database of pharmacokinetics and drug residue literature in fish - 2022 update. *AAPS J.* October 4;24(6):105. http://doi.org/10.1208/s12248-022-00750-w. PMID: 36195686.

della Rocca Ga, ZA, Zanoni R, et al. 2004a. Seabream (*Sparus aurata* L.): Disposition of amoxicillin after single intravenous or oral administration and multiple dose depletion studies. *Aquaculture* 232(1–4):1–10.

Di Salvo A, Pellegrino RM, Cagnardi P, della Rocca G. 2014. Pharmacokinetics and residue depletion of erythromycin in gilthead seabream *Sparus aurata* L. after oral administration. *J Fish Dis* 37(9):797–803.

Doi A, Stoskopf MK, Lewbart GA. 1998. Pharmacokinetics of oxytetracycline in the red pacu (*Colossoma brachypomum*) following different routes of administration. *J Vet Pharmacol Therap* 21(5):364–368.

Fairgrieve WT, Masada CL, McAuley WC, et al. 2005. Accumulation and clearance of orally administered erythromycin and its derivative, azithromycin, in juvenile fall chinook salmon *Oncorhynchus tshawytscha*. *Dis Aquat Organ* 64(2):99–106.

Fairgrieve WT, Masada CL, Peterson ME, et al. 2006. Concentrations of erythromycin and azithromycin in mature chinook salmon *Oncorhynchus tshawytscha* after intraperitoneal injection, and in their progeny. *Dis Aquat Organ* 68(3):227–234.

Ferreira CM, Field CL, Tuttle AD. 2010. Hematological and plasma biochemical parameters of aquarium-maintained cownose rays. *J Aquat Anim Health* 22(2):123–128.

Fleming GJ, Heard DJ, Francis-Floyd R, et al. 2003. Evaluation of propofol and medetomidine-ketamine for short-term immobilization of Gulf of Mexico sturgeon (*Acipenser oxyrinchus de soti*). *J Zoo Wildl Med* 34(2):153–158.

Fleming GJ, Corwin A, McCoy AJ, et al. 2008. Treatment factors influencing the use of recombinant platelet-derived growth factor (Regranex®) for head and lateral line erosion syndrome in ocean surgeonfish (*Acanthurus bahianus*). *J Zoo Wildl Med* 39(2):155–160.

Gaskins J, Boylan S. 2019. Appendix C: Therapeutics for ornamental fish, tropical, bait, and other non-food fish: Supportive therapy and care. In: Smith SA, ed. *Fish Diseases and Medicine*. Boca Raton, FL: CRC Press, 373–377.

GholipourKanani H, Ahadizadeh S. 2013. Use of propofol as an anesthetic and its efficacy on some hematological values of ornamental fish *Carassius auratus*. *SpringerPlus* 2(1):76.

Gratzek JB, Shotts EB, Dawe DL. 1992. Infectious diseases and parasites of freshwater ornamental fish. In: Gratzek JB, Matthews FR, eds. *Aquariology: The Science of Fish Health Management*. Morris Plains, NJ: Tetra Press 227–274.

Greene W, Brookshire G, Delaune AJ. 2018. Hematologic and biochemical summary statistics in aquarium-housed spotted eagle rays (*Aetobatus narinari*). *JZWM* 49(4):912–924.

Greene W, Mylniczenko ND, Storms T, et al. 2020. Pharmacokinetics of ketoprofen in Nile tilapia (*Oreochromis niloticus*) and rainbow trout (*Oncorhynchus mykiss*). *Front Vet Sci* 7:585324.

Grondel JL, Nouws JFM, De Jong M, et al. 1987. Pharmacokinetics and tissue distribution of oxytetracycline in carp, *Cyprinus carpio* L., following different routes of administration. *J Fish Dis* 10(3):153–163.

Guerrero RD. 1975. Use of androgens for the production of all-male *Tilapia aurea* (Steindachner). *Trans Am Fish Soc* 104(2):342–348.

Groff JM, Zinkl JG. 1999. Hematology and clinical chemistry of cyprinid fish. *Vet Clin North Am Exot Anim Pract* 2(3):741–776.

Hansen MK, Nymoen U, Horsberg TE. 2003. Pharmacokinetic and pharmacodynamic properties of Metomidate in turbot (*Scophthalmus maximus*) and halibut (*Hippoglossus hippoglossus*). *J Vet Pharmacol Ther* 26(2):95–103.

Hanson SK, Hill JE, Watson CA, Yanong RPE, Endris R. 2011. Evaluation of emamectin benzoate for the control of experimentally induced infestations of *Argulus* sp. in goldfish and koi carp. *J Aquat Anim Health* 23(1):30–34.

Harms CA. 1996. Treatments for parasitic diseases of aquarium and ornamental fish. *Semin Avian Exot Pet Med* 5(2):54–63.

Harms CA. 1999. Anesthesia in fish. In: Fowler ME, Miller RE, eds. *Zoo and Wild Animal Medicine: Current Therapy 4*. Philadelphia: WB Saunders Co., 158–163.

Harms CA, Lewbart GA. 2000. Surgery in fish. *Vet Clin North Am Exot Anim Pract* 3(3):759–774.

Harms CA, Ross T, Segars A. 2002. Plasma biochemistry reference values of wild bonnethead sharks, *Sphyrna tiburo*. *Vet Clin Pathol* 31(3):111–115.

Harms CA, Lewbart GA, Swanson CR, et al. 2005. Behavioral and clinical pathology changes in koi carp (*Cyprinus carpio*) subjected to anesthesia and surgery with and without intra-operative analgesics. *Comp Med* 55(3):221–226.

Hill JE, Kilgore KH, Pouder DB, et al. 2009. Survey of Ovaprim use as a spawning aid in ornamental fishes in the United States as administered through the University of Florida Tropical Aquaculture Laboratory. *N Am J Aquac* 71(3):206–209.

Homem V, Santos L. 2011. Degradation and removal methods of antibiotics from aqueous matrices – A review. *J Environ Manag* 92(10):2304–2347.

Hrubec TC, Smith SA, Robertson JL, et al. 1996. Blood biochemical reference intervals for sunshine bass (*Morone chrysops* X *Morone saxatilis*) in three culture systems. *Am J Vet Res* 57(5):624–627.

Hrubec TC, Cardinale JL, Smith SA. 2000. Hematology and plasma chemistry reference intervals for cultured tilapia (*Oreochromis* hybrid). *Vet Clin Pathol* 29(1):7–12.

Hrubec TC, Smith SA, Robertson JL. 2001. Age-related changes in hematology and plasma chemistry values of hybrid striped bass (*Morone chrysops* X *Morone saxatilis*). *Vet Clin Path* 30(1):8–15.

https://www.fws.gov/guidance/sites/guidance/files/documents/Erythromycin-study-protocol.pdf. Accessed November 12, 2022.

Johnson EL. 2006. *Koi Health and Disease*. Reade Printers, Athens.

Johnstone R, Macintosh DJ, Wright RS. 1983. Elimination of orally administered 17cr-methyltestosterone by *Oreochromis mossambicus* (tilapia) and *Salmo gairdneri* (rainbow trout) juveniles. *Aquaculture* 35:249–257.

Kilgore KH, Hill JE, Powell JF, Watson CA, Yanong RP. 2009. Investigational use of Metomidate hydrochloride as a shipping additive for two ornamental fishes. *J Aquat Anim Health* 21(3):133–139.

Lewbart GA. 1995. Emergency pet fish medicine. In: Bonagura JD, ed. *Kirk's Current Veterinary Therapy XII: Small Animal Practice*. Philadelphia: WB Saunders Co., 1369–1374.

Lewbart GA. 1998a. Koi medicine and management. *Suppl Comp Contin Educ Pract Vet* 20:5–12.

Lewbart GA. 1998b. Emergency and critical care of fish. *Vet Clin North Am Exot Anim Pract* 1(1):233–249.

Lewbart GA. 1999. CVT update: Antibiotic treatment of aquarium fish. In: Bonagura J, ed. *Current Veterinary Therapy XIII*. Philadelphia: WB Saunders Co., 1196–1198.

Lewbart GA, Butkus DA, Papich M, et al. 2005. A simple catheterization method for systemic administration of drugs to fish. *J Am Vet Med Assoc* 226(5):784–788.

Lewbart GA, Papich MG, Whitt-Smith D. 2005. Pharmacokinetics of florfenicol in the red pacu (*Piaractus brachypomus*) after single dose intramuscular administration. *J Vet Pharmacol Ther* 28(3):317–319.

Lewbart GA, Vaden S, Deen J, et al. 1997. Pharmacokinetics of enrofloxacin in the red pacu (*Colossoma brachypomum*) after intramuscular, oral and bath administration. *J Vet Pharmacol Ther* 20(2):124–128.

Lewbart GA. 2006. Fish supplement. In: Johnson-Delaney C, ed. *Exotic Companion Medicine Handbook*. West Palm Beach, FL: Zoological Medicine Network, 1–58.

Lewis DH, Wenxing W, Ayers A, et al. 1988. Preliminary studies on the use of chloroquine as a systemic chemotherapeutic agent for amyloodinosis in red drum (*Sciaenops ocellatus*). *Mar Sci Suppl* 30:183–189.

Louis MM, Houck EL, Lewbart GA, et al. 2020. Evaluation of potassium chloride administered via three routes for euthanasia of anesthetized koi (Cyprinus carpio). *J Zoo Wildl Med* 51(3):485–489.

McGeachin RB, Robinson EH, Neil WH. 1987. Effect of feeding high levels of androgens on the sex ratio of. *Aquaculture* 61(3–4):317–321.

Minter LJ, Bailey KM, Harms CA, Lewbart GA, Posner LP. 2014. The efficacy of alfaxalone for immersion anesthesia in koi carp (*Cyprinus carpio*). *Vet Anesth Analges* 41(4):398–405.

Morón-Elorza P, Rojo-Solis C, Álvaro-Álvarez T, et al. 2022. Pharmacokinetic studies in elasmobranchs: Meloxicam administered at 0.5 mg/kg using intravenous, intramuscular, and oral routes to nursehound sharks (*Scyliorhinus stellaris*). *Front Vet Sci* 9:845555.

Noga EJ. 2010. *Fish Disease: Diagnosis and Treatment*. 2nd ed. Ames: Wiley-Blackwell.

Oda A, Bailey KM, Lewbart GA, Griffith EH, Posner LP. 2014. Physiologic and biochemical assessments of koi carp, *Cyprinus carpio*, following immersion in propofol. *J Am Vet Med Assoc* 245(11):1286–1291.

Oda A, Messenger KM, Carajal L, et al. 2018. Pharmacokinetics and pharmacodynamic effects in koi carp (*Cyprinus carpio*) following immersion in propofol. *Vet Anaesth Analg* 45(4):529–538.

Ötker HM, Akmehmet-Balcıoğlu I. 2005. Adsorption and degradation of enrofloxacin, a veterinary antibiotic on natural zeolite. *J Hazard Mater* 122(3):251–258.

Otway NM. 2015. Serum biochemical reference intervals for free-living sand tiger sharks (*Carcharias taurus*) from east Australian waters. *Vet Clin Pathol* 44(2):262–274.

Ovaprim (salmon gonadotropin releasing hormone analog 20μg/mL plus domperidone 10 mg/mL) product label. Ferndale, WA: Syndel.

Palmeiro BS, Rosenthal KL, Lewbart GA, et al. 2007. Plasma biochemical reference intervals for koi. *J Am Vet Med Assoc* 230(5):708–712.

Park I. 2019. The anesthetic effects of clove oil and MS-222 on Far Eastern catfish, *Silurus asotus*. *Dev Reprod* 23(2):183–191.

Pattanasiri T, Taparhudee W, Suppakul P. 2017. Acute toxicity and anaesthetic effect of clove oil and eugenol on Siamese fighting fish, *Betta splendens*. *Aquacult Int* 25(1):163–175.

Penning MR, Vaughan DB, Fivaz K, et al. 2017. Chemical immobilization of elasmobranchs at uShaka Sea World in Durban, South Africa. In: Smith M, Warmolts D, Thoney D, et al., eds. *The Elasmobranch Husbandry Manual II: Recent Advances in the Care of Sharks, Rays and Their Relatives*. Columbus, Ohio: Ohio Biological Survey Special Publication of the Ohio Biological Survey 504.

Phu TM, Scippo M, Phuong NT, et al. 2015. Withdrawal time for sulfamethoxazole and trime-thoprim following treatment of striped catfish (*Pangasianodon hypophthalmus*) and hybrid red tilapia (*Oreochromis mossambicus* × *Oreochromis niloticus*). *Aquaculture* 437:256–262.

Plakas SM, DePaola A, Moxey MB. 1991. *Bacillus stearothermophilis* disk assay for determining ampicillin residues in fish muscle. *J Assoc Off Anal Chem Internat* 74:910–912.

Plakas SM, El Said KR, Stehly GR. 1994. Furazolidone disposition after intravascular and oral dosing in the channel catfish. *Xenobiotica* 24(11):1095–1105.

Plakas SM, El Said KR, Bencsath FA, et al. 1998. Pharmacokinetics, tissue distribution and metabolism of acriflavine and proflavine in the channel catfish (*Ictalurus punctatus*). *Xenobiotica* 28(6):605–616.

Plumb JA, Schwedler TE, Limsuwan C. 1983. Experimental anesthesia of three species of freshwater fish with etomidate. *Prog Fish-Cult* 45(1):30–33.

Pottinger TG, Day JG. 1999. A *Saprolegnia parasitica* challenge system for rainbow trout: Assessment of Pyceze as an anti-fungal agent for both fish and ova. *Dis Aquat Organ* 36(2):129–141.

Raulic J, Beaudry F, Beauchamp G, et al. 2021. Pharmacokinetic, pharmacodynamic and toxicology study of robenacoxib in rainbow trout (*Oncorhynchus mykiss*). *J Zoo Wildl Med* 52(2):529–537. doi: 10.1638/2020-0130. PMID: 34130395..

Reimschuessel R, Stewart L, Squibb E, et al. 2005. Fish drug analysis—Phish-Pharm: A searchable database of pharmacokinetics data in fish. *AAPS J* 07(02):E288–327 https://doi.org/10.1208/aapsj070230. http://www.aapsj.org/view.asp?art=aapsj070230. Accessed November 22, 2022.

Roberts HE, Palmeiro B, Weber ES III. 2009. Bacterial and parasitic diseases of pet fish. *Vet Clin North Am Exot Anim Pract* 12(3):609–638.

Roberts HE, Palmeiro B, Weber ES III. 2009. Bacterial and parasitic diseases of pet fish. *Vet Clin North Am Exotic Anim Pract* 12:609–638.

Russo R, Yanong RPE, Mitchell H. 2006. Dietary beta-glucans and nucleotides enhance resistance of red-tail black shark (*Epalzeorhynchos bicolor*, fam. Cyprinidae) to Streptococcus iniae infection. *J World Aquacult Soc* 37(3):298–306.

Russo R, Curtis EW, Yanong RPE. 2007. Preliminary investigations of hydrogen peroxide treatment of selected ornamental fishes and efficacy against external bacteria and parasites in green swordtails. *J Aquat Anim Health* 19(2):121–127.

Saint-Erne N. 1992. Clinical procedures. In: Saint-Erne N, ed. *Advanced Koi Care*. Glendale: Erne Enterprises AZ, 39–61.

Sakamoto K, Lewbart GA, Smith TM II. 2001. Blood chemistry values of juvenile red pacu (*Piaractus brachypomus*). *Vet Clin Pathol* 30(2):50–52.

Samuelsen OB. 2006. Pharmacokinetics of quinolones in fish: A review. *Aquaculture* 255(1–4):55–75.

Scott GL, Law M, Christiansen EF, et al. 2020. Evaluation of localized inflammatory reactions secondary to intramuscular injections of enrofloxacin in striped bass (*Morone saxatilis*). *J Zoo Wildl Med* 51(1):46–52.

Seeley KE, Wolf KN, Bishop MA, Turnquist M, KuKanich B. 2016. Pharmacokinetics of long-acting cefovecin in copper rockfish (*Sebastes caurinus*). *Am J Vet Res* 77(3):260–264.

Siwicki AK, Anderson DP, Rumsey GL. 1994. Dietary intake of immunostimulants by rainbow trout affects non-specific immunity and protection against furunculosis. *Vet Immunol Immunopathol* 41(1–2):125–139.

Skar MW, Haugland GT, Powell MD, et al. 2017. Development of anaesthetic protocols for lumpfish (*Cyclopterus lumpus* L.): Effect of anaesthetic concentrations, sea water temperature and body weight. *PLOS One* 12(7):e0179344.

Steckler, N, Yanong, RP. 2013. *Argulus* (fish louse) infections in fish: FA184, 11/2012. *EDIS* 2013(2) https://doi.org/10.32473/edis-fa184-2012.

Steeil JC, Schumacher J, George RH, et al. 2014. Pharmacokinetics of cefovecin (Convenia ®) in white bamboo sharks (*Chiloscyllium plagiosum*) and Atlantic horseshoe crabs (*Limulus polyphemus*). *J Zoo Wildl Med* 45(2):389–392.

Stoskopf MK. 1999. Fish pharmacotherapeutics. In: Fowler ME, Miller RE, eds. *Zoo and Wild Animal Medicine: Current Therapy 4*. Philadelphia: WB Saunders Co., 182–189.

Stoskopf MK. 1995. Anesthesia of pet fishes. In: Bonagura JD, ed. *Kirk's Current Veterinary Therapy XII: Small Animal Practice*. Philadelphia: Saunders PA, 1365–1369.

Stoskopf MK. 1993. Appendix V: chemotherapeutics. In Stoskopf MK, ed. *Fish Medicine*, 832–839. WB Saunders Co, Philadelphia.

Strobel M, Baker K, Berliner A, et al. 2023. Naltrexone as a promising treatment for clinical signs of lateral line depigmentation in palette surgeonfish (*Paracanthurus hepatus*). *Journal of Zoo and Wildlife Medicine* 54(1): 137–142.

Tavares-Dias M. 2007. Haematological and biochemical reference intervals for farmed channel catfish. *J Fish Biol* 71(2):383–388.

Tocidlowski ME, Lewbart GA, Stoskopf MK. 1997. Hematologic study of red pacu (*Colosomma brachypomum*). *Vet Clin Pathol* 26(3):119–125.

Treves-Brown KM. 2000. *Applied Fish Pharmacology*. Dodrecht, The Netherlands: Kluwer Academic Publishers.

Tripathi NK, Latimer KS, Brunley VV. 2004. Hematologic reference intervals for koi (*Cyprinus carpio*), including blood cell morphology, cytochemistry, and ultrasturcture. *Vet Clin Pathol* 33(2):74–83.

Varadaraj K, Pandian TJ. 1989. Monosex male broods of *Oreochromis mossambicus* produced through artificial sex reversal with 17α-methyl-4 androsten-17α-ol-3-one. *Curr Trends Life Sci* 15:169–173.

Vorbach BS, Chandasana H, Derendorf H, et al. 2019. Pharmacokinetics of oxytetracycline in the giant danio (*Devario aequipinnatus*) following bath immersion. *Aquaculture* 498:12–16.

Waxman L, Fustukjian A, Thompson V. 2021. A better way to buffer: Pilot study on buffering capability of sodium bicarbonate and sodium carbonate used in conjunction with tricaine methanesulfonate (MS-222) during anesthetic procedures in marine and freshwater settings. *Proc Int Assoc Aquat Anim Med (virtual)*.

Whaley J, Francis-Floyd R. 1991. A comparison of metronidazole treatments of hexamitiasis in angelfish. *Proc Int Assoc Aquat Anim Med* 1991: 110–114.

Whitaker BR. 1999. Preventive medicine programs for fish. In: Fowler ME, Miller RE, eds. *Zoo and Wild Animal Medicine: Current Therapy 4*. Philadelphia: WB Saunders Co., 163–181.

Whitehead M, Vanetten CL, Zheng Y, Lewbart GA. 2019. Hematological parameters in largemouth bass (*Micropterus salmoides*) with formalin-preservation: Comparison between wild tournament-caught and captive-raised fish. *PeerJ*. https://doi.org/10.7717/peerj.6669.

Wildgoose WH, Lewbart GA. 2001. Therapeutics. In: Wildgoose WH, ed. *Manual of Ornamental Fish*. 2nd ed. Gloucester: British Small Animal Veterinary Association 237–258.

Wright SE, Stacy NI, Yanong RPE,, et al. 2021. Hematology and biochemistry panel reference intervals for captive saddleback *Amphiprion polymnus* and tomato clownfish *Amphiprion frenatus*. *J Aquat Anim Health* 33(1):3–16.

Xu W, Zhu X, Wang X, et al. 2006. Residues of enrofloxacin, furazolidone and their metabolites in Nile tilapia (*Oreochromis niloticus*). *Aquaculture* 254(1–4):1–8.

Yanong RPE. 2021. Preliminary investigations into use of metomidate for euthanasia of ornamental fishes. *J Aquat Anim Health* 33(3):133–138.

Yanong RPE, Curtis EW. 2005. Pharmacokinetic studies of florfenicol in koi carp and threespot gourami *Trichogaster trichopterus* after oral and intramuscular treatment. *J Aquat Anim Health* 17:129–137.

Yanong RPE. 2012. unpublished data.

Yanong RPE, Russo R, Curtis E, et al. 2004. *Cryptobia iubilans* infection in juvenile discus (*Symphysodon aequifasciata*): Four case reports, pathology, and treatment trials. *J Am Vet Med Assoc* 224(10):1644–1650.

Zimmerman DM, Armstrong DL, Curro TG, et al. 2006. Pharmacokinetics of florfenicol after a single intramuscular dose in white-spotted bamboo sharks (*Chiloscyllium plagiosum*). *J Zoo Wildl Med* 37(2):165–173.

CHAPTER 15
EQUIPMENT AND SUPPLIES CHECKLIST

HUSBANDRY

- Air pumps.
- Air tubing.
- Airline valves.
- Airstones.
- Assorted plastic totes/sweater boxes.
- Assorted glass aquaria with covers.
- Assorted sizes of plastic fish bags.
- Assorted nets (various sizes; also include a koi net).
- Commercial dechlorinator.
- Hides (shelters) – aquarium safe clay pots, PVC pipes, plastic plants.
- Rubber bands.
- Salt – sea salt and sodium chloride.
- Siphon hoses.
- Sponge filters.
- Professional water quality test kit.
    - TAN, nitrite, pH, alkalinity, hardness, thermometer/temperature.
    - Nitrate, chloride, salinity (salt refractometer and/or hydrometer).
    - Chlorine.
- Dissolved oxygen meter.
- Water sample bottles (plastic, 250 ml).
- 5-gallon bucket(s).

MEDICAL SUPPLIES

- Centrifuge.
- Complete dissecting kit.
- Compound microscope.
- Eugenol (clove oil) 1:9 with 95% ethanol (stock approx. 100 mg/ml).
- Fish anesthesia delivery system (FADS) machine.

DOI: 10.1201/9781003057727-15

- Gram scale (to 1 kg).
- Kg scale (to 100 kg).
- MS-222 (10 mg/ml buffered stock solution; buffered with sodium bicarbonate or sodium carbonate).
- Oxygen tank with regulator.
- Plastic surgical drapes.
- Refractometer.
- Sterile surgical pack(s).

MICROBIOLOGY

- Culturettes (if culture growth/analysis to be run at outside lab).
- Glass slides (for impression smears).
- Benchtop acid fast bacteria stain.
- Acid fast bacillli (AFB) control slide.
- Gram stain.
- If primary/initial cultures to be run in house:
  - 28°C incubator (or, if room temperature is 23-28°C, can incubate on benchtop (or in a Tupperware® container)).
  - Tryptic soy agar with 5% sheep's blood (blood agar) plates (good general media).
  - Brain heart infusion broth (for incubating blood cultures prior to plating).
  - Ethanol (90%–95%) and alcohol burner (for flame disinfection of instruments).
  - Inoculating loops and wires (reusable metal) or disposable (may have less yield than metal).
  - Antibiotic sensitivity disks and Mueller Hinton agar.

DRUGS AND OTHER COMPOUNDS

(See Chapter 14 for additional drugs/compounds and dosing)

- Amikacin.
- Assorted scalpel blades.
- Atropine.
- Calcium gluconate.
- Ceftazidime.

- Commercial dechlorinator.
- Copper sulfate (pentahydrate; "blue" copper) or an OTC aquarium chelated copper product.
- Dexamethasone injectable.
- Diflubenzuron.
- Emeraid II.
- Enrofloxacin 22.7 mg/ml injectable.
- Enrofloxacin 22.7 mg tablets.
- Epinephrine.
- Eugenol (clove oil) 1:9 with 95% ethanol (stock approx. 100 mg/ml).
- Euthanasia solution.
- Fenbendazole.
- Florfenicol.
- Formalin (100%); OTC products available.
- Furosemide.
- Heparin 1000 u.
- Relevant, fresh fish food appropriate for the species (Although not ideal, commercial Oxbow Critical Care® (carnivore/omnivore/herbivore) or similar may serve temporarily).
- Lidocaine 2%.
- Lubricating jelly.
- Lufenuron.
- Metronidazole.
- MS-222 (10 mg/ml buffered stock solution).
- 10% Neutral buffered formalin.
- Nexaband® adhesive.
- Nitrofurazone.
- Nolvasan® (dilute and nondilute).
- Oxytetracycline.
- Panalog® ointment or another similar product.
- Potassium permanganate.
- Povidone iodine ointment.
- Praziquantel tablets and injectable solution.
- Sea salt (5 lbs (2.25 kg)).
- Silver sulfadiazine cream.
- Sterile saline.
- Sterile water.
- Trichlorfon 8%.
- Trimethoprim/sulfamethoxazole tablets 960 mg.

DIAGNOSTIC SUPPLIES

- Assorted needles (0.5-1.5 inch): 18g, 20g, 22g, 23g, 25g, 26g.
- Assorted red rubber catheters 5 French–12 French.
- Assorted Silastic™ laboratory tubing of various inner diameter (ID)/outer diameter (OD; including one of ID 1.57 mm and OD 2.41 mm) for use in ovarian biopsy of koi. ID should be large enough for mature oocytes, and OD is sized for the gonoduct leading into the ovary. Smaller sizes may be needed based on size of fish and species.
- Assorted suture materials.
- Assorted syringes: 1 cc, 3 cc, 6 cc, 12 cc, 35 cc, 60 cc with lure or eccentric tip.
- 2.5 ml blood collection tubes red top (clot tube).
- Diff-Quik type stain.
- Green (heparin).
- Yellow (sodium chloride).
- Bullet tubes (eppendorf) 1.5 ml.
- Camera (with microscope attachment or cell phone photo/recording)
- Centrifuge(s).
- Coverslips.
- Culturettes.
- Cutting/necropsy board.
- Drill and drill bits (for thicker, older animals – blood culture).
- Minitip portacult and regular ARD bottle.
- Fluorescein strips.
- Forceps (various).
- Glass slides.
- Isopropyl alcohol (70%) and alcohol wipes for surface sterilization of skin prior to culture.
- Jars for histology.
- Kimwipes®.
- Markers, pens, pencils.
- Microcentrifuge with putty.
- Microhematocrit tubes.
- Microcontainer tubes (serum separator).
- Neutral buffered formalin (10%) and jars for histopathology samples.
- Nonsterile 2x2 and 4x4 pads.
- Nonsterile gloves.
- Paper towels.
- Parafilm.

- Plastic pipettes.
- Paper or digital template for recording necropsy highlights.
- RNA Later (or other preservative for PCR; can also freeze tissue samples).
- Ruler.
- Scale(s).
- Scalpel handles and blades (various sizes).
- Scissors (various sizes).
- Sample jars.
- Tongue depressors.
- Trump's fixative (for electron microscopy).
- Whirlpak bags.
- Ziploc/sealable plastic bags.

CHAPTER 16

MATERIAL RESOURCES

PRODUCTS AND SERVICES

Aqueon Products*
(aquarium tanks, filters/equipment, accessories, diets)
Central Aquatics
5401 West Oakwood Park Drive
Franklin, WI 53132
888-255-4527
https://www.aqueon.com/

Coralife*
(aquarium tanks, filters/equipment, accessories)
800-255-4527
Email: info@central-aquatics.com
https://www.coralifeproducts.com/

Kent Marine*
(water conditioning products, nutritional additives, sea salt, supplies)
https://www.kentmarine.com/

  * A division of Central Garden & Pet Company
  1340 Treat Blvd, Suite 600
  Walnut Creek, CA 94597

AquaClear**
(aquarium filters/equipment)
http://usa.hagen.com/aquaclear

Fluval**
(aquarium tanks, filters/equipment, accessories, diets)
https://fluvalaquatics.com/us/

DOI: 10.1201/9781003057727-16

**A division of Rolf C. Hagen (USA) Corp.
305 Forbes Blvd. Mansfield,
MA, 02048
Customer Service: 1-800-724-2436
Dealer Support: 1-800-353-3444
Dealers: http://www.hagendirect.com
Website: http://usa.hagen.com

Instant Ocean***
(artificial sea salt, some supplies)
https://www.instantocean.com/

Marineland***
(aquarium tanks, filters/equipment, accessories, diets)
https://www.marineland.com/

OMEGAONE***
(fish food, diets)
https://www.omegasea.net/ 1-888-204-3273

Tetra***
(aquarium tanks, filters/equipment, accessories, diets)
https://www.tetra-fish.com/

***A division of Spectrum Brands Pet, LLC
3001 Commerce St., Blacksburg, VA 24060-6671
800-526-0650

WATER CHEMISTRY TEST KITS

Hach Company
PO Box 389, Loveland CO 80539
800-227-4224
www.hach.com

La Motte Chemicals
PO Box 329
Chestertown MD 21260

800-344-3100
www.lamotte.com

YSI Inc.
1700/1725 Brannum Lane
Yellow Springs, OH 45387
877-726-0975 (US)
https://www.ysi.com/

FISH DRUG DISTRIBUTORS

- Veterinary drug supplier preferred, but some relevant drugs may be found at local pet stores and other aquarium/pet supply retailers

OTHER INFORMATIONAL RESOURCES

National Fish/Aquatic Animal Health Organizations
- American Association of Fish Veterinarians
  - fishvets.org
- World Aquatic Veterinary Medical Association
  - wavma.org
- International Association for Aquatic Animal Medicine (IAAAM)
  - www.iaaam.org
- American Fisheries Society – Fish Health Section
  - https://units.fisheries.org/fhs/

Industry Resources
- American Cichlid Association
  - www.cichlid.org
- American Pet Products Association
  - https://www.americanpetproducts.org/
- Associated Koi Clubs of America
  - www.akca.org
- Florida Tropical Fish Farmers Association (FTFFA)
  - www.ftffa.com
- National Ornamental Goldfish Growers Association (NOGGA)
  - 6916 Black's Mill Rd, Thurmont, MD 21788
  - Phone: (301)271-7475

- Ornamental Fish International (OFI)
  - www.ofish.org
- Pet Industry Joint Advisory Council (PIJAC)
  - www.pijac.org
- University of Florida IFAS Tropical Aquaculture Laboratory
  - https://tal.ifas.ufl.edu/
- UF IFAS TAL Extension Publications
  - https://tal.ifas.ufl.edu/extension-and-outreach/extension-publications/

COURSES RELEVANT FOR PET FISH MEDICINE

- University of Florida (School of Forest, Fisheries, and Geomatics Sciences, and College of Veterinary Medicine): Online Diseases of Warmwater Fish and Aquaculture and Fish Health Graduate Certificate Program
- University of Georgia short course on koi medicine
- AQUAVET® (Cornell University)
- E-quarist Course ® (The Aquarium Vet)
- Marvet

POPULAR MAGAZINES

- Amazonas Magazine
  - https://www.amazonasmagazine.com/

AQUATIC MEDIA PRESS, LLC
- 3075 Rosemary Ln NE, Rochester, MN, 55906-4535
- 800-217-3523

Coral Magazine
- https://coralmagazine.com/

REEF TO RAINFOREST MEDIA, LLC
- 490 Acorn Lane, P.O. Box 490, Shelburne, VT 05482
- (802) 985-9977

Tropical Fish Hobbyist
- 1 TFH Plaza Neptune City NJ 07753
- 908-988-8400
- www.tfhmagazine.com

REFERENCES/FURTHER READING

Anderson ET, Stoskopf MK, Morris Jr JA, Clarke EO, Harms CA. 2010. Hematology, plasma biochemistry, and tissue enzyme activities of invasive red lionfish captured off North Carolina, USA. *J Aquat Anim Health* December 1;22(4):266–273.

Arnold JE. 2005. Hematology of the sandbar shark, Carcharhinus plumbeus: Standardization of complete blood count techniques for elasmobranches. *Veterinary Clinical Pathology* 34(2):115–123.

Boylan S. 2011, September. Zoonoses associated with fish. *Veterinary Clinics of North America: Exotic Animal Practice* 14(3): 427–438.

Brown L. 1993. Anesthesia and restraint. In Stoskopf MK, ed. *Fish Medicine*, 81. WB Saunders.

Cain DK, Harms CA, Segars A. 2004. Plasma biochemistry reference values of wild-caught southern stingrays (*Dasyatis americana*). *J Zoo Wildl Med* 35(4):471–476.

Campbell TW, Ellis C. 2007. Hematology of fish. In *Avian & Exotic Animal Hematology & Cytology*. Blackwell Publishing, Ames, IA, 93–111.

Clark TS, Pandolfo LM, Marshall CM, Mitra AK, Schech JM. 2018. Body condition scoring for adult Zebrafish (Danio rerio). *J Am Assoc Lab Anim Sci*, October 25;57(6):698–702. doi: 10.30802/AALAS-JAALAS-18-000045. Epub ahead of print. PMID: 30360771; PMCID: PMC6241379.

Collins S, Dornburg A, Flores JM, Dombrowski DS, Lewbart GA. 2016. A comparison of blood gases, biochemistry, and hematology to ecomorphology in a health assessment of pinfish (*Lagodon rhomboides*) *PeerJ* 4:e2262. https://doi.org/10.7717/peerj.2262.

Ferguson HW, Bjerkas E, Evensen O. 2006. *Systemic Pathology of Fish: A Text and Atlas of Normal Tissue Responses in Teleosts, and Their Responses in Disease*. Scotian Press, Ames, IA, 368 pp.

Francis-Floyd R, Watson C, Petty D, Pouder D. 2022. Ammonia in aquatic systems: FA-16/FA031, 06/2022. *EDIS* 2022(4):1–6.

Gratzek JB. 1994. *Aquariology: The Science of Fish Health Management*. Tetra Press, Morris Plains, NJ, 330 pp.

Greene W, Brookshire G, Delaune AJ. 2018. Hematologic and biochemical summary statistics in aquarium-housed spotted eagle rays (*Aetobatus narinari*). *J Zoo Wildl Med* December;49(4):912–24.

Grier HJ. 2012. Development of the follicle complex and oocyte staging in red drum, *Sciaenops ocellatus* Linnaeus, 1776 (Perciformes, Sciaenidae). *Journal of Morphology* 273(8): 801–829.

Grier HJ, Uribe-Aranzábal MC, Patiño R. 2009. The ovary, folliculogenesis, and oogenesis in teleosts. *Reproductive Biology and Phylogeny of Fishes (Agnathans and Bony Fishes)* 8(Part A): 25–84.

Groff JM, Zinkl JG. 1999. Hematology and clinical chemistry of cyprinid fish. *Vet Clin North Am: Exotic Anim Pract* 2:741–776 (Carassius auratus). *J Zoo Wild Med* 48(3): 776–785.

Hadfield C, Clayton L. 2022. *Clinical Guide to Fish Medicine*. Wiley-Blackwell, Hoboken, New Jersey, 610 pp.

Harms C, Ross T, Segars A. 2002. Plasma biochemistry reference values of wild bonnethead sharks, Sphyrna tiburo. *Vet Clin Pathol* September;31(3):111–115.

Harms C. 2005. Surgery in fish research: Common procedures and postoperative care. *Lab Animal* 34(1): 28–34.

Hoffman GL. 1999. *Parasites of North American Freshwater Fishes*. Cornell University Press, Ithaca.

Hrubec TC, Cardinale JL, Smith SA. 2000. Hematology and plasma chemistry reference intervals for cultured tilapia (*Oreochromis* hybrid). *Vet Clin Path* 29(1): 7–12.

Hrubec TC, Smith SA, Robertson JL. 2001. Age-related changes in hematology and plasma chemistry values of hybrid striped bass (Morone chrysops X Morone saxatilis). *Vet Clin Path* 30(1): 8–15.

Johnson EJ. 1997. *Koi Health and Disease*. Johnson Veterinary Services, Marietta, GA, 141 pp.

Katz EM, Chu DK, Casey KM, Jampachaisri K, Felt SA, Pacharinsak C. 2020. The stability and efficacy of tricaine methanesulfonate (MS222) solution after long-term storage. *Journal of the American Association for Laboratory Animal Science* 59(4): 393–400.

Klinger R, Francis-Floyd R, Riggs A, Reed P. 2003. Use of blood culture as a nonlethal method for isolating bacteria from fish. *Journal of Zoo and Wildlife Medicine* 34(2): 206–207.

Lewbart GA. 1998. *Self-Assessment Color Review of Ornamental Fish*. Iowa State University Press, Ames, Iowa, 192 pp.

Lewbart GA. 2017. *Self-Assessment Color Review of Ornamental Fishes and Aquatic Invertebrates*. CRC Press, Boca Raton, Florida, 247 pp.

Lewbart GA, Stone EA, Love NE. 1995. Pneumocystectomy in a Midas cichlid. *J Am Vet Med Assoc* 207(3):319–321.

Lewbart GA, Harms CA. 1999. Building a fish anesthesia delivery system. *Exotic DVM Magazine* 1(2):25–28.

Longshaw M, Feist SW. 2001. Parasitic diseases. In Wildgoose WH, ed. *BSAVA Manual of Ornamental Fish*, 2nd ed., 167–183 British Small Animal Veterinary Association, Gloucester.

Lowry T, Smith SA. 2007. Aquatic zoonoses associated with food, bait, ornamental and tropical fish. *Journal of the American Veterinary Medical Association* 231(6): 876–880.

Masser MP, Jensen JW. 1991. *Calculating Area and Volume of Ponds and Tanks*. USDA: Southern Regional Aquaculture Center.

Noga EJ. 2010. *Fish Disease: Diagnosis and Treatment*, 2nd ed. Wiley-Blackwell, Ames, Iowa, 536 pp.

Noga EJ, Wang C, Grindem CB, Avtalion R. 1999. Comparative clinicopathological responses of striped bass and palmetto bass to acute stress. *Trans Am Fish Soc* 128:680–686.

Palmeiro BS, Rosenthal KL, Lewbart GA, et al. 2007. Plasma biochemical reference intervals for koi. *J Am Vet Med Assoc* 230:708–712.

Plumb JA, Hanson KA. 2011. *Health Maintenance and Principal Microbial Diseases of Cultured Fishes*, 3rd ed. Wiley-Blackwell, Ames, Iowa, 506 pp.

Preena PG, Kumar VJR, Singh ISB. 2021. Nitrification and denitrification in recirculating aquaculture systems: The processes and the players. *Reviews in Aquaculture* 13: 2053–2075.

Roberts RJ. 2012. *Fish Pathology*, 4th ed. Wiley, Ames, Iowa, 592 pp.

Roberts HE (ed.). 2009. *Fundamentals of Ornamental Fish Health*. Wiley-Blackwell, Ames, IA, 244 pp.

Ross LG. 2001. Restraint, anaesthesia, and euthanasia. In Wildgoose WH, ed. *BSAVA Manual of Ornamental Fish*, 2nd ed., 76. BSAVA, Gloucester.

Saint-Erne N. 2010. *Advanced Koi Care*, 2nd ed. Erne Enterprises, Glendale, AZ, 194 pp.

Saint-Erne N. 2019. In experimental use of metomidate anesthesia for ornamental fish euthanasia. *The Aquatic Veterinarian* 13(1):32–33.

Sakamoto K, Lewbart GA, Smith TA, II. 2001. Blood chemistry values of juvenile red pacu (*Piaractus brachypomus*). *Vet Clin Pathol* 30(2):50–52.

Sanders J. 2023. *How to Kill Your Koi*. Google Books, 136 pp.

Smith SA. 2019. *Fish Diseases & Medicine*. CRC Press, Boca Raton, Florida, 396 pp.

Smith SA, Harms CA. 2023. Fish. In *Carpenter's Exotic Animal Formulary*, 6th ed. Elsevier, St. Louis, Missouri, 22–71.

Spotte S. 1992. *Captive Seawater Fishes*. Lea and Febiger, New York, 942 pp.

Stoskopf MK. 1993. *Fish Medicine*. W.B. Saunders Co., Philadelphia, PA, 882 pp.

Stoskopf MK. 2010. *Fish Medicine* (Second Printing). ART Sciences. Available for $99 on www.Lulu.com.

The Merck Veterinary Manual, 11th ed. Merck & Co., Inc., National Publishing Co., Philadelphia, PA, 2016. https://www.merckvetmanual.com/.

Tocidlowski ME, Lewbart GA, Stoskopf MK. 1997. Hematologic study of red pacu (*Colosomma brachypomum*). *Vet Clin Pathol* 26(3):119–125.

Treves-Brown KM. 2000. *Applied Fish Pharmacology*. Kluwer Academic Publishers, Dordrecht, The Netherlands.

Tripathi NK, Latimer KS, Brunley VV. 2004. Hematologic reference intervals for koi (*Cyprinus carpio*), including blood cell morphology, cytochemistry, and ultrasturcture. *Vet Clin Pathol* 33:74–83.

Westmoreland LSH, Archibald KE, Christiansen EF, Broadhurst HJ, Stoskopf MK. 2019. The mesopterygial vein: A reliable venipuncture site for intravascular access in batoids. *Journal of Zoo and Wildlife Medicine* 50(2): 369–374.

Wildgoose WH (ed.). 2001. *BSAVA Manual of Ornamental Fish*, 2nd ed. Gloucester.

Woo PTK, ed. 2006. *Fish Diseases and Disorders, Volume 1: Protozoan and Metazoan Infections*, 2nd ed. CAB International, Cambridge, MA.

Woo PTK, Bruno DW, eds. 2011. *FIsh Diseases and Disorders, volume 3: Viral, Bacterial, and Fungal Infections*, 2nd ed. CAB International, Cambridge, MA.

Wright SE, Stacy NI, Yanong RPE, Juhl RN, Lewbart GA. 2021. Hematology and biochemistry panel reference intervals for captive saddleback *Amphiprion polymnus* and tomato clownfish *Amphiprion frenatus*. *J Aquat Anim Health* 33:3–16.

Whitehead M, Vanetten CL, Zheng Y, Lewbart GA. 2019. Hematological parameters in largemouth bass (*Micropterus salmoides*) with formalin-preservation: Comparison between wild tournament-caught and captive-raised fish. *PeerJ*. https://doi.org/10.7717/peerj.6669.

Yanong RPE. 1996. Reproductive management of freshwater ornamental fish. *Seminars in Avian and Exotic Pet Medicine* 5(4): 222–235.

Yanong RPE. 1999. Nutrition of ornamental fish. In Jenkins JR, ed. *Husbandry and Nutrition*. W.B. Saunders Co., Philadelphia, PA. *Veterinary Clinics of North America: Exotic Animal Practice* 2(1): 19–42.

Yanong RPE. 2003. Fungal diseases of fish. In Jones MP, ed. *Fungal Diseases*. W.B. Saunders Co., Philadelphia, PA. *Veterinary Clinics of North America, Exotic Animal Practice* 6(2): 377–400.

Yanong RPE, Ehrlacher-Reid C. 2012. Biosecurity in aquaculture, part 1: An overview. USDA Southern Regional Aquaculture Center Publication No. 4707.

Yanong RPE, Martinez CV, Watson CA. 2009. Use of Ovaprim in ornamental fish aquaculture. FA 161. Program in Fisheries and Aquatic Sciences, School of Forest, Fisheries, and Geomatics Sciences, Florida Cooperative Extension Service, Institute of Food and Agricultural Sciences, University of Florida, Gainesville. http://edis.ifas.ufl.edu/fa161.

INDEX

A

Acidity, 56
Adult zebrafish, 80, 81
Aeration, 40
Aeromonas salmonicida, 199
Aeromoniasis, 221
Aggressive aquarium species, 2
Albino bristlenose catfish (*Ancistrus temminckii*), 7, 8
Alkalinity, 56
Ambiphyra sp., 168
Ammonia, 58
Ammonia oxidizing archaea (AOA), 53
Ammonia oxidizing bacteria (AOB), 53
Amyloodinium, 177, 241, 242
Anchorworm (*Lernaea* sp.), 195
Anchorworm parasites, 221
Anesthesia
  fish formulary, 244–249
  stages in fishes, 151
  tips, 152
Angelfish (*Pterophyllum scalare*), 14, 31, 97, 189
  common species, 32
  french angelfish (*Pomacanthus paru*), 31, 32
Animal & Plant Health Inspection Service (APHIS), 220
Ante-mortem diagnostics
  bacterial culture, 102–106
  blood collection techniques
    bulbus/cardiac draw, 108, 109
    cardiac draw, 107–108
    caudal vein/vessels, 107
    cut tail/peduncle draw, 108, 109
    elasmobranch, 108, 110–112
    gill arch vessels, 107
  diagnostic evaluation, 113
  external wet mount biopsies, 102, 103
  hematology, 113–122
AOA, *see* Ammonia oxidizing archaea
AOB, *see* Ammonia oxidizing bacteria
Aphanomyces invadans, 218, 220
Aquaculture systems
  biofiltration, 38
    fixed film biofilter, 36, 40
    nitrification biofiltration cycle, 37, 39
    thick *vs.* thin film, 36, 39
  circulation, 36
  filtration
    in parallel circuit, 38
    in series circuit, 37
  major processes, 37
  solids capture, 36
  solids types, 38
Aquarium non-fish species, 34
Argulus sp., 194
Asian tapeworm (*Bothriocephalus acheilognathi*), 184
Atlantic blue tang (*Acanthurus coeruleus*), 163
Autogenous vaccine, 160

B

Bacterial diseases, 166, 201
Banggai cardinalfish (*Pterapogon kauderni*), 22
BCS, *see* Body condition scoring
Bettas, 87–88
Bicarbonate, 57
Bioballs filter, 43, 46
Biofiltration, 36, 38
  fixed film biofilter, 36, 40

nitrification biofiltration cycle, 37, 39
  thick vs. thin film, 36, 39
Bipolar cautery, 157
Black ghost knifefish (*Apteronotus albifrons*), 183
Blennies
  common species, 24
  Kamohara blenny (*Meiacanthus kamoharai*), 23
  striped blenny (*Meiacanthus grammistes*), 23
Blood collection techniques
  bulbus/cardiac draw, 108, 109
  cardiac draw, 107–108
  caudal vein/vessels, 107
  cut tail/peduncle draw, 108, 109
  elasmobranch, 108, 110–112
  gill arch vessels, 107
Blue paradise fish (*Macropodus opercularis*), 16
Blue ram dwarf cichlid (*Mikrogeophagus ramirezi*), 12, 13
Body condition scoring (BCS), 77, 80, 81, 227
Bony fish, 147, 148
Bottom-sitting, 228
Bowfin (*Amia calva*), 126
Brain culture, 134, 135
Brine shrimp, 79, 80
Brooklynella sp., 166
Bubble bead filter, 43, 45
Bulbus/cardiac draw, 108, 109
Buoyancy problem, 222, 224
Butterflyfishes
  common species, 28
  copperband butterflyfish (*Chelmon rostratus*), 27
Butterfly koi (*Cyprinus rubrofuscus*), 124

C

Calcium, 58
Cannister power filter, 43, 46
Capillaria pterophylli, 189
Capriniana sp., 170
Carbohydrates, 77
Carbon dioxide, 38, 40, 56, 140, 244
Cardiac draw, 107–108

Cardinalfish
  Banggai cardinalfish (*Pterapogon kauderni*), 22
  common species, 23
  pajama cardinalfish (*Sphaeramia nematoptera*), 22
Cardinal tetra (*Paracheirodon axelrodi*), 179
Carps and barbs
  comet goldfish (*Carassius auratus*), 4, 5
  common species, 7
  goldfish, 6
  juvenile koi (*Cyprinus rubrofuscus*), 4, 6
  koi fish (*Cyprinus rubrofuscus*), 6
  long-finned zebra danio, 4
  rainbow shark (*Epalzeorhynchos frenatum*), 4, 5
  redcap oranda goldfish (*Carassius auratus*), 4, 6
  redtail black shark (*Epalzeorhynchos bicolor*), 4, 5
  rosy barb (*Pethia conchonius*), 5
Catfishes
  albino bristlenose catfish (*Ancistrus temminckii*), 7, 8
  common species, 8–9
  peppered cory cat (*Corydoras paleatus*), 7
  pleco (*Hypostomus plecostomus*), 7
Caudal vein/vessels, 107
Cestode larva, 174
Channel catfish (*Ictalurus punctatus*), 191
Chilodonella sp., 165
Chlorine/bleach, 52
Chlorine/chloramine toxicity, 70
Chlorinity, 70
Chloroquine, 161, 175, 241
Cichlids, 85–87, 90, 91, 133
  angelfish, 14
  blue ram dwarf cichlid (*Mikrogeophagus ramirezi*), 12, 13
  common species, 15–16
  dwarf cichlids, 14, 16
  dwarf South American cichlids, 15, 16
  freshwater angelfish (*Pterophyllum scalare*), 12, 13
  Lake Tanganyika cichlid (*Neolamprologus brichardi*), 12, 13
  new world cichlids, 14, 15
  old world cichlids, 14, 15

INDEX

oscar cichlid (*Astronotus ocellatus*), 12, 14
peacock cichlid (*Aulonocara* sp.), 11, 12
Circling, 228
Clark's clownfish (*Amphiprion clarkii*)., 19, 20
Cleaner wrasse (*Labroides dimidiatus*), 28
Clinical visit
  annual examination protocol
    history questions, 95–96
    site visit, 95
    with virtual site visit, 95
    water quality management, 97
  diagnostics, 98, 100
  emergency therapy/first aid, 100
  endoscopic evaluation, 100
  physical examination, 97–98
  tube feeding, 101
Clove oil, 150, 152, 245, 246
Clownfish (*Amphiprion ocellaris*), 212–214, 222–223
  Clark's clownfish (*Amphiprion clarkii*)., 19, 20
  common species, 20–21
  false clownfish (*Amphiprion ocellaris*), 19
Clown loach (*Chromobotia macracanthus*), 9
Clown triggerfish (*Balistoides conspicillum*), 29
Columnaris, 206, 208, 209, 232, 234, 235, 237
Comet goldfish (*Carassius auratus*), 4, 5
Constipation, 228
Copper, 72
Copperband butterflyfish (*Chelmon rostratus*), 27
Corneal opacity, 227
Courses, pet fish medicine, 270
Crustacean parasites, 163
Cryptobia iubilans, 146, 147, 172–174
Cryptocaryon irritans, 163
Cut tail/peduncle draw, 108, 109

D

Damselfish, 19
  common species, 20–21
  yellowtail damselfish (*Chrysiptera parasema*), 20

Deficiency signs
  minerals, 81, 83–84
  vitamins, 81, 82
Degasification, 40
Denitrifying bacteria, 53
Dermocystidium, 176, 179
Diagnostic supplies, 265–266
Diatomaceous earth (DE) filter, 46
Digenean, 160, 181–182, 187–188, 243
Dinoflagellate, 160, 175, 177
Disinfection, 40
Doppler pencil probe, 156, 157
Dorsal approach, 135
Dottybacks
  common species, 22
  neon dottyback (*Pseudochromis aldabraensis*), 21
  yellow dottyback (*Pseudochromis fuscus*), 21
Dragonets, 26–27
Drugs, 263–264
Drugs and chemicals
  and dosages, 233–237, 240
  fish formulary, 239–240
    anti-fungal agents, 233–237
    anti-microbial agents, 233–237
    anti-parasitic agents, 243
    sedation, anesthesia, and euthanasia, 244–249
  for ornamental fish, 232, 238–239
  pre-treatment considerations, 232
Dwarf cichlids, 14, 16
Dwarf gourami (*Trichogaster lalius*)., 16

E

Edwardsiella, 221
Egg binding, 92–94
Egg bound, 85
Egg layers, 90–91
Elasmobranch, 108, 110–112
Emperor tetra (*Nematobrycon palmeri*), 2, 3
Epistylis sp., 168
Epizootic ulcerative syndrome (EUS), 220
Equipment and supplies
  diagnostic supplies, 265–266
  drugs, 263–264
  husbandry, 262

medical supplies, 262–263
microbiology, 263
Ergasilus sp., 195, 196
Erysipelothrix, 202–203
Ethanol, 150
Eugenol, 150, 152, 155, 245, 246, 262, 264
Eustrongylides (nematode) worms, 189
Euthanasia
    fish formulary, 244–249
    methods, 149–150
        immersion methods, 150
        injectable methods, 150–151
Exophthalmia, 226–227
External wet mount biopsies, 102, 103

F

FADS, *see* Fish anesthesia delivery system
False clownfish (*Amphiprion ocellaris*), 19
Families and species
    aggressive aquarium species, 2
    angelfish, 31
        common species, 32
        French angelfish (*Pomacanthus paru*), 31, 32
    aquarium non-fish species, 34
    blennies
        common species, 24
        Kamohara blenny (*Meiacanthus kamoharai*), 23
        striped blenny (*Meiacanthus grammistes*), 23
    butterflyfishes
        common species, 28
        copperband butterflyfish (*Chelmon rostratus*), 27
    cardinalfish
        Banggai cardinalfish (*Pterapogon kauderni*), 22
        common species, 23
        pajama cardinalfish (*Sphaeramia nematoptera*), 22
    carps and barbs
        comet goldfish (*Carassius auratus*), 4, 5
        common species, 7
        goldfish (*Carassius auratus*), 6
        juvenile koi (*Cyprinus rubrofuscus*), 4, 6
        koi fish, 6
        long-finned zebra danio, 4
        rainbow shark (*Epalzeorhynchos frenatum*), 4, 5
        redcap oranda goldfish (*Carassius auratus*), 4, 6
        redtail black shark (*Epalzeorhynchos bicolor*), 4, 5
        rosy barb (*Pethia conchonius*), 5
    catfishes
        albino bristlenose catfish (*Ancistrus temminckii*), 7, 8
        common species, 8–9
        peppered cory cat (*Corydoras paleatus*), 7
        pleco (*Hypostomus plecostomus*), 7
    cichlids
        angelfish, 14
        blue ram dwarf cichlid (*Mikrogeophagus ramirezi*), 12, 13
        common species, 15–16
        dwarf cichlids, 14, 16
        dwarf South American cichlids, 15, 16
        freshwater angelfish (*Pterophyllum scalare*), 12, 13
        Lake Tanganyika cichlid (*Neolamprologus brichardi*), 12, 13
        new world cichlids, 14, 15
        old world cichlids, 14, 15
        oscar cichlid (*Astronotus ocellatus*), 12, 14
        peacock cichlid (*Aulonocara* sp.), 11, 12
    clownfish
        Clark's clownfish (*Amphiprion clarkii*)., 19, 20
        common species, 20–21
        false clownfish (*Amphiprion ocellaris*), 19
    common tetra species, 4
    damselfish, 19
        common species, 20–21
        yellowtail damselfish (*Chrysiptera parasema*), 20

Index

dottybacks
  common species, 22
  neon dottyback (*Pseudochromis aldabraensis*), 21
  yellow dottyback (*Pseudochromis fuscus*), 21
dragonets, 26–27
gobies
  common species, 26
  Mandarin goby (*Synchirops splendidus*), 27
  neon goby (*Elacatinus oceanops*), 25
  watchman goby (*Cryptocentrus cinctus*), 25, 26
groupers and basslets
  clown triggerfish (*Balistoides conspicillum*), 29
  common species, 31
  panther grouper (*Cromilepetes altivelis*), 29, 30
  royal gramma (Gramma loreto), 29, 30
  Swalesi basslet (*Liopropoma swalesi*), 29, 30
hawkfish, 24
labyrinth fish
  blue paradise fish (*Macropodus opercularis*), 16
  common species, 17
  dwarf gourami (*Trichogaster lalius*), 16
livebearers
  common species, 12
  gold dust molly (*Poecilia* sp.), 10, 11
  male guppy (*Poecilia reticulata*), 10, 11
  male sunburst platy (*Xiphophorus* sp.), 10, 11
  male swordtail (*Xiphophorus hellerii*), 10, 11
loaches
  clown loach (*Chromobotia macracanthus*), 9
  common species, 10
  Pakistani loach (*Botia almorhae/lohachata*), 9, 10
  zebra loaches (*Botia striata*), 9, 10

mature cardinal tetra, 2
mature neon tetra, 2
pond fish, 18
puffers, 25
rainbowfish
  common species, 18
  red rainbowfish (*Glossolepis incisus*), 18
  turquoise rainbowfish (*Melanotaenia lacustris*), 17, 18
single pet aquarium species, 2
tangs and surgeonfish, 32
  common species, 34
  juvenile aquacultured yellow tang (*Zebrasoma flavescens*), 32, 33
  Pacific blue tang (*Paracanthurus hepatus*), 32, 33
  yellow tang (*Zebrasoma flavescens*), 32, 33
triggerfish, 29
wrasses
  cleaner wrasse (*Labroides dimidiatus*), 28
  common species, 29
Fatty liver disease, 222
Firemouth cichlid (*Thorichthys meeki*), 143–146
Fish anesthesia delivery system (FADS), 153–155
Fish drug distributors, 269
Fish formulary, 239–240
  anti-fungal agents, 233–237
  anti-microbial agents, 233–237
  anti-parasitic agent, 243
  sedation, anesthesia, and euthanasia, 244–249
Fixed film biofilter, 36, 40
Flame hawkfish (*Neocirrhites armatus*), 24
Flavobacterium columnare, 133
Foam fractionator, 45, 48
Food and Drug Administration (FDA), 239
Formalin, 56, 93, 102, 113, 138, 161–162, 171, 181, 215, 235, 242, 243, 264, 265
French angelfish (*Pomacanthus paru*), 31, 32

Freshwater angelfish (*Pterophyllum scalare*), 12, 13
Freshwater livebearers, 92
Freshwater tropical egg layers, 92
Full-spectrum light, 43
Fungal diseases, 214–215, 217
Fusariomycosis, 216, 219

G

Gas bubbles, 224
Genital papilla, 85
Gill arch vessels, 107
Gobies
   common species, 26
   mandarin goby (*Synchirops splendidus*), 27
   neon goby (*Elacatinus oceanops*), 25
   watchman goby (*Cryptocentrus cinctus*), 25, 26
Gold dust molly (*Poecilia* sp.), 10, 11
Goldenback triggerfish, 78, 79
Goldfish (*Carassius auratus*), 6, 75, 78, 81, 88–89, 133, 136, 140
   viral diseases, 211–212
Gonadal tumors, 222, 223
Gourami (*Trichopodus trichopterus*), 87–88, 147
Granulomas, 201
Grey mullet (*Mugil cephalus*), 218
Groupers and basslets
   clown triggerfish (*Balistoides conspicillum*), 29
   common species, 31
   panther grouper (*Cromilepetes altivelis*), 29, 30
   royal gramma (Gramma loreto), 29, 30
   Swalesi basslet (*Liopropoma swalesi*), 29, 30
Gulping, 229

H

Halfmoon betta (*Betta splendens*), 87
Hawkfish, 24
Head and lateral line erosion (HLLE) syndrome, 225

Health certificates, 230–231
Hematology, 113–122
Hemorrhage, 227
Heteropolaria sp., 169
Hospital tank disinfection, 159
Hovering, 228
Hydrogen peroxide, 52

I

Ichthybodo sp., 173
Ichthyophonus, 176, 180
Ichthyophthirius multifiliis (Ich) organisms, 163
Imaging
   radiography, 123–126
   ultrasonography, 126
Immersion methods, 150
Infectious disease
   bacterial diseases, 166–167, 201–202
   carp edema virus, 207, 209
   clinical signs, 197–198
   epitheliocystis, 197, 202, 204–205
   fungal diseases, 214–215, 217
   hospital tank disinfection, 159
   koi herpesvirus, 220
   koi pox, 205, 208
   lymphocystis, 98, 113, 205, 211, 213–214
   Megalocytivirus, 211, 213
   microbiome, 159
   nitrifying bacteria, 159
   parasitic diseases, 160–165
   spring viremia of carp, 220
   vaccines, 160
   viral diseases, 205–206
      goldfish, 206–207
      koi, 206–207
      parasites, 161–167
   zoonoses, 220–221
Injectable methods, 150–151
Internal evaluation, 136–140
Intracoelomic (ICe) injection, 238
Intramuscular injection, 238
Intravenous injection, 238
Isopod, 160, 193, 196
Isopropyl alcohol, 134, 135

J

Juvenile aquacultured yellow tang (*Zebrasoma flavescens*), 32, 33
Juvenile koi (*Cyprinus rubrofuscus*), 4, 6
Juvenile tilapia (*Oreochromis* sp.), 204

K

Kamohara blenny (*Meiacanthus kamoharai*), 23
Ketamine-medetomidine, 150
Koi (*Cyprinus rubrofuscus*), 6, 78, 88–89, 124, 133, 136, 137, 155, 194, 209–210
  viral diseases, 211–212
Koi herpesvirus (KHV), 220
Koi pox, 205, 208

L

Labyrinth fish, 90
  blue paradise fish (*Macropodus opercularis*), 16
  common species, 17
  dwarf gourami (*Trichogaster lalius*), 16
Lake Tanganyika cichlid (*Neolamprologus brichardi*), 12, 13
Lateral line depigmentation syndrome (LLD), 225
Livebearers, 87, 91
  common species, 12
  gold dust molly (*Poecilia* sp.), 10, 11
  male guppy (*Poecilia reticulata*), 10, 11
  male sunburst platy (*Xiphophorus* sp.), 10, 11
  male swordtail (*Xiphophorus hellerii*), 10, 11
Loaches
  clown loach (*Chromobotia macracanthus*), 9
  common species, 10
  Pakistani loach (*Botia almorhae/lohachata*), 9, 10
  zebra loaches (*Botia striata*), 9, 10
Long-finned zebra danio, 4

M

Macronutrients, 77
Magnesium, 58
Malachite green, 194, 204, 236
Male guppy (*Poecilia reticulata*), 10, 11
Male sunburst platy (*Xiphophorus* sp.), 10, 11
Male swordtail (*Xiphophorus hellerii*), 10, 11
Mandarin goby (*Synchirops splendidus*), 27
Marine sharks, 147
Material resources
  courses, pet fish medicine, 270
  fish drug distributors, 269
  industry resources, 269–270
  magazines, 270
  National Fish/Aquatic Animal Health Organizations, 269
  products and services, 267–268
  water chemistry test kits, 268–269
Mature cardinal tetra, 2
Mature neon tetra, 2
Medical supplies, 262–263
Mesomycetozoea, 214
Metazoan parasites, 160
Metomidate hydrochloride, 150
Microbiological evaluation, 134–136
Microbiology, 263
Microbiology supplies, 128–129
Microbiome, 159
Micronutrients, 78
Microsporidian, 178
Milletseed butterflyfish (*Chaetodon miliaris*), 226
Minor Use and Minor Species (MUMS) Animal Health Act, 240
Mirror carp (*Cyprinus rubrofuscus*), 209
Monogenean, 75, 132, 160, 184–186, 241–243
Moribund fish, 129
MS-222, 155
Multi-cellular metazoan parasites, 160
Mycobacterium, 221
Myxozoan, 197

N

National Fish/Aquatic Animal Health Organizations, 269
National List of Reportable Animal Diseases (NLRAD), 220
National Veterinary Accreditation Program, 230
Necropsy, 133
Necropsy supplies, 128
Neighbor syndrome, 81
Nematode parasite, 189
Neobenedenia sp., 185–186
Neon dottyback (*Pseudochromis aldabraensis*), 21
Neon tetra (*Paracheirodon innesi*), 145
Neoplasia, 222
Nessler's reagent, 57
New world cichlids, 14
Nitrate, 70
Nitrification biofiltration cycle, 37, 39
Nitrifying bacteria, 159
Nitrite, 69, 71, 73
Nitrite oxidizing bacteria (NOB), 53
Nitrogen cycle, 36, 53–55
NOB, *see* Nitrite oxidizing bacteria
Normal fish anatomy, 140–148
Nuptial tubercles, 88, 89
Nutrient leaching, 78
Nutrition, 90
    macronutrients, 77
    micronutrients, 78
    nutrient leaching, 78
    requirements, 77
    storing flakes, 78
Nutritional diseases, 227–228

O

Ocular diseases, 226–227
Office International des Epizooties (OIE), 220
Old world cichlids, 14, 15
One-celled protistan parasites, 160
Organophosphate, 181–182, 250
Oscar cichlid (*Astronotus ocellatus*), 12, 14, 136, 138, 139, 205
Oxygen, 38, 40, 42, 49–51, 53, 55, 71, 73, 95, 140, 224, 253, 262, 263
Ozonation, 41
Ozone, 41

P

Pacific blue tang (*Paracanthurus hepatus*), 32, 33, 225
Packed cell volume (PCV), 113
Pajama cardinalfish (*Sphaeramia nematoptera*), 22
Pakistani loach (*Botia almorhae/lohachata*) and zebra loaches (*Botia striata*), 9, 10
Panther grouper (*Cromilepetes altivelis*), 29, 30
PAR, *see* Photosynthetically active radiation
Parasites, 160
Parasitic diseases, 160–164
Parasitic turbellarian, 188
Parrotfish (*Scarus coelestinus*), 219
PCV, *see* Packed cell volume
Peacock bass (*Cichla* sp.), 125
Peacock cichlid (*Aulonocara* sp.), 11, 12
Pentastome nymph, 190
Pentobarbital, 150
Peppered cory cat (*Corydoras paleatus*), 7
pH, 53–58, 71
Photosynthetically active radiation (PAR), 43Piping, 224
Piscinoodinium sp., 177
Pleco (*Hypostomus plecostomus*), 7
Plecostomus species, 2
Pleistophora, 175, 178
Polydioxanone suture, 158
Polyopisthocotylean, 186–187
Pond fish, 1, 18
Ponds, 74–75, 225
Post-mortem evaluation
    internal evaluation, 136–140
    microbiological evaluation, 134–136
    microbiology supplies, 128–129
    necropsy, 133
        supplies, 128
    normal fish anatomy, 140–148
    specimen quality, 127

wet mount evaluation, 129–133
Potassium permanganate, 193, 195, 237, 243, 264
Propofol, 151
Protandrous hermaphrodites, 91
Protogynous hermaphrodites, 91
Protoopalinid protozoan, 174
Puffers, 25

Q

Quaternary ammonium compounds (QACs), 52

R

Radiography, 123–126
Rainbowfish, 88
    common species, 18
    red rainbowfish (*Glossolepis incisus*), 18
    turquoise rainbowfish (*Melanotaenia lacustris*), 17, 18
Rainbow shark (*Epalzeorhynchos frenatum*), 4, 5
Red blood cell (RBCs), 113
Redcap oranda goldfish (*Carassius auratus*), 4, 6
Red dovii cichlid (*Parachromis dovii*), 85
Red rainbowfish (*Glossolepis incisus*), 18
Red Sea bream iridoviral disease (RSIVD), 219–220
Redtail black shark (*Epalzeorhynchos bicolor*), 4, 5
Reef systems, 75
Reproduction
    bettas, 87–88
    cichlids, 85–87
    egg binding, 92–94
    egg bound, 85
    goldfish, 88–89
    gouramis, 87–88
    koi, 88–89
    livebearers, 87
    rainbowfishes, 88
    reproductive strategies
        egg layers, 90–91
        freshwater species, 89–90

        parental care, 90–91
Rosy barb (*Pethia conchonius*), 5
Royal gramma (Gramma loreto), 29, 30

S

Salinity, 74
Salmincola sp., 195
Salmonella, 221
Saltwater tips, 74
Saprolegnia, 214, 216, 217, 234, 235, 242
Seahorse (*Hippocampus* sp.), 141, 142
Sedation, 151–152
    fish formulary, 244–249
Serpae tetra (*Hyphessobrycon eques*), 2, 3
Sexing/sex, 6, 17, 19, 85–89, 91, 96, 142, 143, 153, 171–172, 250, 252
Showa koi (*Cyprinus rubrofuscus*), 126
Silver dollars (*Metynnis* sp.), 2, 3
Single pet aquarium species, 2
Sodium bicarbonate, 57
Sodium chloride (NaCl), 75
Solar salt, 75
Spadefish (*Chaetodipterus faber*), 125
Spanish mackerel (*Scomberomorus maculatus*), 196
Spironucleus vortens, 173
Sponge filter, 44–45
Spotted green puffer (*Dichotomyctere nigroviridis*), 25
Spring viremia of carp (SVC), 220
Storing flakes, 78
Streptococcus, 221
Striped blenny (*Meiacanthus grammistes*), 23
Subcutaneous/intradermal injection, 238
Submersible heater, 49
Superficial wounds, 224
Surgery
    anesthetic stage/plane, 155
    common indications, 153
    doppler pencil probe, 153, 156
    fish anesthesia delivery system, 153–154
    MS-222, 155
    supplies and equipment, 156
    tips, 157–158

Swalesi basslet (*Liopropoma swalesi*), 29, 30
Swim bladder, 148, 157
Swordtail (*Xiphophorus helleri*), 87
Synchronous hermaphrodites, 91
Synodontis catfish, 222, 223

T

TAN, *see* Total ammonia nitrogen
Tangs and surgeonfish, 32
   common species, 34
   juvenile aquacultured yellow tang (*Zebrasoma flavescens*), 32, 33
   Pacific blue tang (*Paracanthurus hepatus*), 32, 33
   yellow tang (*Zebrasoma flavescens*), 32, 33
Tank essentials
   adsorption, 46
   cover, 43
   filter
      bioballs filter, 43, 46
      biofiltration/nitrification, 43
      bubble bead filter, 43, 45
      cannister power filter, 43, 46
      diatomaceous earth filter, 43, 46
      foam fractionator, 47, 48
      hang-on aquarium back power filter, 43
      mechanical filtration, 43
      sponge filter, 43, 44
      ultraviolet sterilizer filters, 48
   light, 42
   nets, 52
   photosynthetically active radiation, 42
   siphon hose, 51
   size, 40–41
   submersible heater, 49
Tank tips, 72
   new biofilter, 71–72
   new tank, 71
Tetrahymena, 162, 165
Thick film biofilters, 39
Thin film biofilters, 39
Thrombocytes, 113
Tiger barbs (*Puntius tetrazona*), 203
Tilapia, leeches in, 191, 204

Total ammonia nitrogen (TAN), 53, 55
Tricaine methanesulfonate, 150, 151
Trichodina/trichodinid, 102, 161, 243
Triggerfish, 29
Tube feeding, 101
turbellarian, 188
Turquoise rainbowfish (*Melanotaenia lacustris*), 17, 18

U

UIA, *see* Un-ionized ammonia
Ultrasonography, 126
Ultraviolet (UV) sterilization, 40
Ultraviolet sterilizer filter, 48–49
Un-ionized ammonia (UIA), 42, 53–54
Upside down swimming, 229
Uronema, 162
USDA APHIS VS, 230
US Department of Agriculture's (USDA), 220, 230, 231

V

Vaccines, 160
Veterinary Export Health Certification System (VEHCS), 230
Vibrio, 221
Viral diseases, 205–206
   goldfish, 206–207
   koi, 206–207
   parasites
      metazoan–crustacea and myxozoa, 192–193
      metazoan–worm/worm-like, 161, 181
      protists–flagellates, 171–172
      protists–motile ciliates, 161–162
      protists–sessile ciliates, 167
   tropical aquarium fish, 211
Virkon, 52

W

Watchman goby (*Cryptocentrus cinctus*), 25, 26
Water

chemistry test kits, 268–269
quality, 90
    acidity, 56
    alkalinity, 56
    ammonia, 58
    calcium, 58
    carbon dioxide, 38, 40, 56, 140, 244
    chlorine/chloramine toxicity, 70
    copper, 70
    gravity to salinity conversion table, 59–60
    hardness, 58
    nitrate, 70
    nitrite, 69, 71
    oxygen, 38, 40, 42, 49–51, 53, 55, 70–71, 73, 95, 140, 224, 253, 262, 263
    pH, 56–58
salt calculations, 75
tips, 74–75
WBCs, *see* White blood cells
Wet mount evaluation, 129–133, 160

White blood cells (WBCs), 113
World Organization for Animal Health (WOAH), 219
Worm-like parasites, 160
Wrasses
    cleaner wrasse (*Labroides dimidiatus*), 28
    common species, 29

Y

Yellow dottyback (*Pseudochromis fuscus*), 21
Yellowtail damselfish (*Chrysiptera parasema*), 20
Yellow tang (*Zebrasoma flavescens*), 32, 33

Z

Zap dose, 41
Zebra danio (*Danio rerio*), 99, 203
Zebra loaches (*Botia striata*), 9, 10
Zoonoses, 220–221

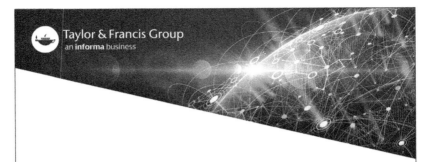

For Product Safety Concerns and Information please contact our EU representative GPSR@taylorandfrancis.com Taylor & Francis Verlag GmbH, Kaufingerstraße 24, 80331 München, Germany

Printed and bound by CPI Group (UK) Ltd, Croydon, CR0 4YY
08/06/2025
01896985-0002